Lecture Notes in Control and Information Sciences

Edited by A.V. Balakrishnan and M. Thoma

66

Systems and Optimization

Proceedings of the Twente Workshop
Enschede, The Netherlands, April 16–18, 1984

Edited by
A. Bagchi and H. Th. Jongen

Springer-Verlag
Berlin Heidelberg New York Tokyo

Editors
Arunabha Bagchi
Hubertus Theodorus Jongen

Department of Applied Mathematics
Twente University of Technology
P..O. Box 217
7500 AE Enschede
The Netherlands

ISBN 3-540-15004-8 Springer-Verlag Berlin Heidelberg New York Tokyo
ISBN 0-387-15004-8 Springer-Verlag New York Heidelberg Berlin Tokyo

Offsetprinting: Mercedes-Druck, Berlin
Binding: Lüderitz und Bauer, Berlin
2061/3020-543210

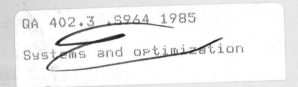

PREFACE

Modern control theory and nonlinear programming originated independently and almost simultaneously during the middle fifties. After a decade or so, control problems were successfully formulated in the nonlinear programming framework in appropriate function spaces and Pontryagin's maximum principle was derived from the Kuhn-Tucker Theory. Dynamic programming, in the meanwhile, had a far reaching impact in both optimization and control theory. From the early seventies, however, the two disciplines clearly started drifting apart as control theorists got increasingly interested in the system-theoretic aspects of dynamical models. Although this state of affairs is, to some extent, understandable, a total divorce of these two disciplines would be most unfortunate. The purpose of the workshop on Systems and Optimization held on April 16-18, 1984, here at the Twente University of Technology was to bring the mainly Dutch scientists in these areas together with some renowned international experts in order to stimulate interaction among researchers working in systems and in optimization. We hope that the publication of this book, which consists of lectures presented by the invited speakers during the workshop, will serve similar purpose on a larger scale.

The workshop was organized by the participants of the research program Systems and Optimization initiated last year at the Twente University of Technology. The objective of this program is to investigate the theoretical and practical aspects of systems theory and optimization, but more importantly, to study the inter-relations between these two disciplines. The topics covered inevitably reflect a slight bias of the researchers organizing the workshop and there is a clear concentration of lectures on parametric optimization and nonlinear systems theory. Otherwise, quite a large number of topics have been covered.

The workshop was made possible through generous financial support from the Netherlands Organization for the Advancement of Pure Research (ZWO) and the Hogeschoolfonds Twente Foundation. Our sincere thanks go to both of these organizations for their support. We take this opportunity to thank all our colleagues in the research program Systems and Optimization for their active interest and advice. Finally, our thanks to Ms. Carla Hassing-Assink and Ms. Marja Langkamp for cheerfully shouldering all secretarial responsibilities and for editorial assistance in preparing this volume.

October, 1984. Arunabha Bagchi

Enschede, The Netherlands. Bert Jongen

4186

LIST OF PARTICIPANTS

In the following list lecturers are indicated by an asterisk.

AXELSSON, J.P.,
Lund Institute of Technology, Division of Automatic Control, S-220 07 Lund 7 — Sweden

BAGCHI, A.,
Twente University of Technology, Dept. of Applied Math. — Netherlands

*BASAR, T.,
University of Illinois, Coordinated Science Laboratory, Urbana, Illinois 61801 — USA

BOEKHOUDT, P.,
Twente University of Technology, Dept. of Applied Math. — Netherlands

BOSGRA, O.H.,
Delft University of Technology, Dept. of Mechanical Eng. Mekelweg 2, 2628 CD Delft — Netherlands

*BROSOWSKI, B.,
Joh. Wolfg. Goethe-Universität, Fachbereich Mathematik, Robert-Mayer-Strasse 6-10, D-6000 Frankfurt a.M. — F.R.G.

*CROUCH, P.E.,
University of Warwick, Control Theory Centre, Coventry CV4 7AL, Warwickshire — U.K.

CURTAIN, R.,
University of Groningen, Dept. of Mathematics, P.O. Box 800, 9700 AV Groningen — Netherlands

*DONTCHEV, A.L.,
Bulgarian Academy of Sciences, Inst. Mathematics with Computer Centre, P.O. Box 373, 1090 Sofia — Bulgaria

ESSEN VAN, G.J.,
Hollandse Signaalapparaten BV, P.O. Box 42, 7550 GD Hengelo — Netherlands

*FIACCO, A.V.,

The George Washington University, Dept. of Operations Research Washington, D.C. 20052 — USA

*FRANCIS, B.A.,

University of Toronto, Dept. of Electrical Eng., Toronto, Ontario M5S 1A4 — Canada

*GUDDAT, J.,

Humboldt Universität zu Berlin, Sektion Mathematik, Unter den Linden 6, PSF 1297, DDR-1086 Berlin — G.D.R.

HANZON, B.,

Delft University of Technology, Dept. of Mathematics and Computer Science, P.O. Box 356, 2600 AJ Delft — Netherlands

*HETTICH, R.P.,

Universität Trier, Fb. 4 Math., Postfach 3825, D-5500 Trier — F.R.G.

HOVE TEN, D.,

Twente University of Technology, Dept. of Applied Math. — Netherlands

*ISIDORI, A.,

Università di Roma, Instituto di Automatica, Via Eudossiana 18, 00184 Roma — Italy

JONGEN, H.Th.,

Twente University of Technology, Dept. of Applied Math. — Netherlands

JONKER, P.,

Twente University of Technology, Dept. of Applied Math. — Netherlands

*KNOBLOCH, H.W.,

Universität Würzburg, Mathematisches Institut der Universität, Am Hubland, 8700 Würzburg — F.R.G.

KUILDER, H.,

Hollandse Signaalapparaten BV, P.O. Box 42, 7550 GD Hengelo — Netherlands

KWAKERNAAK, H.,

Twente University of Technology, Dept. of Applied Math. — Netherlands

LILJA, M.,
Lund Institute of Technology, Division of Automatic Control, S-220 07 Lund 7 — Sweden

MAARSEVEEN, M.F.A.M.,
IWIS-TNO, Schoemakerstraat 97, 2628 VK Delft — Netherlands

MERBIS, M.,
University of Tilburg, Department of Economics, P.O. Box 90153, 5000 LE Tilburg — Netherlands

NIJMEIJER, H.,
Twente University of Technology, Dept. of Applied Math. — Netherlands

OLSDER, G.J.,
Delft University of Technology, Dept. of Mathematics and Computer Science, P.O. Box 356, 2600 AJ Delft — Netherlands

*PARDOUX, E.,
Université de Provence, UER de Mathématique, 3 Place V. Hugo, 13331 Marseille Cedex 3 — France

POLDERMAN, J.W.,
Center for Mathematics and Computer Science, P.O. Box 4079, 1009 AB Amsterdam — Netherlands

SCHAFT VAN DER, A.J.,
Twente University of Technology, Dept. of Applied Math. — Netherlands

SCHUMACHER, J.M.,
Center for Mathematics and Computer Science, P.O. Box 4079, 1009 AB Amsterdam — Netherlands

SCHUPPEN VAN, J.H.,
Center for Mathematics and Computer Science, P.O. Box 4079, 1009 AB Amsterdam — Netherlands

SCHUT, J.,
Twente University of Technology, Dept. of Applied Math. — Netherlands

SMULDERS, S.,
Twente University of Technology, Dept. of Applied Math. Netherlands

SPREIJ, P.J.C.,
Center for Mathematics and Computer Science, P.O. Box 4079, 1009 AB Amsterdam Netherlands

STRIJBOS, R.C.W.,
Twente University of Technology, Dept. of Applied Math. Netherlands

SIJGERS, I.,
Twente University of Technology, Dept. of Applied Math. Netherlands

TRENTELMAN, A.L.,
University of Groningen, Dept of Mathematics, P.O. Box 800, 9700 AV Groningen Netherlands

TWILT, F.,
Twente University of Technology, Dept. of Applied Math. Netherlands

*VARAIYA, P.,
University of California at Berkeley, Dept. of Electrical Eng. and Computer Science, Berkeley, CA 94720 USA

VERMEULEN, E.,
Twente University of Technology, Dept. of Applied Math. Netherlands

*WAGNER, K.,
Universität Würzburg, Mathematisches Institut der Universität, Am Hubland, 8700 Würzburg F.R.G.

WETTERLING, W.W.E.,
Twente University of Technology, Dept. of Applied Math. Netherlands

WILLEMS, J.C.,
University of Groningen, Dept. of Mathematics, P.O. Box 800, 9700 AV Groningen Netherlands

ZWIER, G.,
Twente University of Technology, Dept. of Applied Math. Netherlands

CONTENTS

DYNAMIC GAMES AND INCENTIVES

Tamer Başar
Decision and Control Laboratory
Department of Electrical and Computer Engineering
and Coordinated Science Laboratory
University of Illinois
1101 W. Springfield Avenue
Urbana, Illinois 61801, USA

1. Introduction

In this paper we present a general mathematical formulation and a method of solution for stochastic incentive decision problems, using concepts and tools of dynamic game theory. An incentive problem involves at least two decision makers, with the decision law (policy) of one of them structurally affecting the cost (or utility) function of the other. This functional dependence enables the dominant decision maker to dictate certain actions on the other decision maker, by using smooth (not threat) policies. As discussed in the paper, these problems could be viewed as nonzero-sum dynamic games under the Nash or Stackelberg solution concepts. After the mathematical formulation, these game theoretic concepts are introduced in Section 2, together with a discussion of different approaches that could be undertaken in case of incomplete description of the problem (such as one decision maker not knowing the cost function of the other decision maker precisely).

Section 3 presents four different types of incentive decision problems which can be viewed as special cases of the general formulation. The last one introduced in this section is then taken up in Section 4 and investigated in some detail, together with a sensitivity analysis in case of unknown parameters. The paper concludes with the illustrative example of Section 5.

2. Problem Formulation and Motivation

The basic ingredients of a stochastic incentive decision problem are the following:

There are at least two decision makers (DM's), with at least two levels of hierarchy in decision making, the top level being occupied by a single decision maker called the "leader." The leader declares incentives, and the other DM's, called the "follower(s)," act based on these declared incentives. For the sake of the discussion to follow, and without losing much generality of conceptual nature, we assume that there is only one follower (F) with decision variable $v \in V$, where V is an appropriate (finite or infinite dimensional) space. The leader (L)'s decision variable, on the other hand, will be denoted by $u \in U$, where U is his decision space. The notation "ξ" will denote the "random state of Nature" which, at this point of generality, could be a random variable, a random vector or a stochastic process defined on an underlying probability space.

The leader and the follower will be allowed to make some (possibly imperfect) measurement of ξ, which we denote by $z_L(v,\xi)$ and $z_F(\xi)$ for L and F, respectively. The asymmetry in the roles of L and F shows itself here, in the sense that z_L may depend explicitly on the act of F, which L may or may not observe directly. Hence, the information available to L will be z_L and some (perfect or imperfect) version of v, which we denote as $\eta_L(v,z_L)$. The follower (F), on the other hand, will have access to z_F, as well as the announced incentive policy of L, which is denoted by γ; we denote this information for F by $\eta_F(\gamma,z_F) = (\gamma,z_F)$ which is clearly an invertible mapping.

The admissible policies of L and F are, respectively, $\gamma:\eta_L \to u$, $\beta:\eta_F \to v$, belonging to some appropriate spaces Γ_L and Γ_F. The former will also be called the *incentive policy* of L. Note that for each fixed pair of policies $(\gamma,\beta)\in\Gamma_L \times \Gamma_F$, the actions of L and F are determined by the "closed-loop" relations

$$\left.\begin{aligned} u &= \gamma(\eta_L) = \gamma(\eta_L(v,z_L(v,\xi))) \\ v &= \beta(\eta_F) = \beta(\gamma,z_F(\xi)) \end{aligned}\right\} \tag{1}$$

which are uniquely solvable, leading to a pair $(u(\xi),v(\xi))$.[†]

A common stipulation in incentive design problems is to view F as a rational DM whose goal is to optimize (under the given information η_F) a known objective functional which quantifies his goals. Let us take this objective functional as the expected value of a loss functional to be minimized:

$$J_F(\gamma,\beta) = E\{L_F(u = \gamma(\eta_L), v = \beta(\eta_F),\xi)\} \tag{2}$$

where the expectation is over the prior statistics of ξ. Hence, F's optimal actions are determined by

$$v^o_\gamma = \beta^o(\eta_F) = \beta^o(\gamma,z_F) \tag{3}$$

where β^o minimizes J_F over Γ_F. In (3) we use γ as a subscript in v to bring to attention the fact that v^o_γ can also be viewed as the optimal response of F to the announced incentive policy γ of L. Knowing the structure of this mapping, L designs a policy γ that would eventually lead to an outcome which is desirable by him. Note that since v^o depends explicitly on γ, L can "control" the structure of the optimization problem faced by F and can therefore force him "in principle" to a desired action. This desired action could be an element out of $\mathfrak{V}$, defined as the set of z_F-measurable random variables whose realizations are in V. Let us denote such a choice by

$$v^d \in \mathfrak{V} \tag{4a}$$

which is in general a stochastic varialbe leading, under the particular policy γ that enforces this action, to

[†]Here, we have abused notation and have used $u(\xi)$, $v(\xi)$ to denote random variables (<u>not</u> points in U and V); but their realizations do belong to the decision spaces U and V.

$$u^d = \gamma(\eta_L(v^d, z_L(v^d,\xi))) \in \mathfrak{U}\ ^\dagger \tag{4b}$$

which is also stipulated to be a desired act on part of L. Now, if there exists no $\gamma \in \Gamma_L$ which would induce the desired $v^d \in \mathfrak{V}$, a "best" incentive policy for L would be one that minimizes an appropriate norm of the difference between v^d and v^o_γ, i.e.

$$\| v^o_\gamma - v^d \| \rightarrow \text{minimize over } \gamma \in \Gamma_L \quad . \tag{5}$$

A second type of an objective for L would be the (global) minimization of the expected value of a particular loss functional,

$$J_L(\gamma,\beta) = E\{L_L(u = \gamma(\eta_L),\ v = \beta(\eta_F), \xi)\} \quad , \tag{6a}$$

by taking into account the fact that β is determined by (3). Hence, L's problem is then

$$J_L(\gamma, \beta^o(\gamma,\cdot)) \rightarrow \text{minimize over } \gamma \in \Gamma_L \quad , \tag{6b}$$

which provides him with the best incentive policy γ that achieves the desired outcome.

Equilibrium Solution

To provide a precise definition for the (equilibrium) solution of this incentive design problem, we view it as a two-person game in normal form, described by cost functionals J_L and J_F, where J_L could also be replaced by $\| v - v^d \|$ if v^d is the ultimate desired act for F. Then, an incentive policy $\gamma^o \in \Gamma_L$ is called an optimal (or Stackelberg) policy if

$$J_L(\gamma^o, \beta^o_{\gamma^o}) = \min_{\gamma \in \Gamma_L} J_L(\gamma, \beta^o_\gamma) \tag{7a}$$

where β^o_γ satisfies for each $\gamma \in \Gamma_L$

$$J_F(\gamma, \beta^o_\gamma) = \min_{\beta \in \Gamma_F} J_F(\gamma,\beta) \quad . \tag{7b}$$

Note that in this case it is not important that F have access to γ through his information structure (i.e. η_F may not involve γ), because the enforcement of γ on F is provided through the solution concept, by virtue of the presence of hierarchy in decision making.

Another possible equilibrium solution that has been adopted for such incentive problems in the literature is the Nash solution which is symmetric (as a solution concept) and makes explicit use of the availability of γ to F through his information structure. In this case we will say that a pair $(\gamma^N, \beta^N) \in \Gamma_L \times \Gamma_F$ is in Nash equilibrium if

---

† The set $\mathfrak{U}$ is the set of $\gamma_L(v^d, z_L(v^d,\xi))$-measurable stochastic variables taking values in U.

$$J_L(\gamma^N,\beta^N) = \min_{\gamma\in\Gamma_L} J_L(\gamma,\beta^N(\gamma,z_F)) \tag{8a}$$

$$J_F(\gamma^N,\beta^N) = \min_{\beta\in\Gamma_F} J_F(\gamma^N,\beta(\gamma^N,z_F)) \tag{8b}$$

Note that in our context every Stackelberg equilibrium is also a Nash equilibrium, but not vice versa because (7b) is a stronger requirement than (8b). While a number of papers in the economics literature have adopted the Nash equilibrium concept for such problems, we are going to work here with the Stackelberg solution which can also be viewed as a "strong Nash equilibrium" stripped off any informational nonuniqueness [Başar and Haurie, 1984].

Additional Elements that Contribute to Uncertainty

It is not always realistic to assume that L knows perfectly the rational response β_γ of F to his announced policies γ. This may arise, for example, if L does not know J_F (or rather the loss function L_F) either totally or to within some parametric values. The latter is more interesting for our purposes, in which case we can write L_F as

$$L_F = L_F(u,v,\xi,\alpha) \tag{9}$$

where α represents the parameter (or parameter vector) in F's loss functional, whose value is known by him but not by L. Because of the presence of this unknown parameter α, L does not know (or cannot compute) F's optimal response policy β_γ, since (in general) it explicitly depends on the value of α; β_γ^α. This renders the minimization problem faced by L meaningless unless he makes some rational assumption concerning the exact value or the distribution of possible values of α. One such approach is the "Bayesian" one adopted by Harsanyi (1968) where L develops a subjective probability distribution for α; Harsanyi shows that such a probability distribution can always be determined, provided that one takes as given the validity of a number of axioms. A second approach is the so-called "minimax design" where L chooses a policy which minimizes his objective functional under worst possible choice(s) of α; i.e. L solves

$$\min_{\gamma\in\Gamma_L} \max_{\alpha\in A} J_L(\gamma,\beta_\gamma^\alpha) \tag{10}$$

where A is the a priori chosen set in which possible values of α lie, and β_γ^α is F's optimal response, explicitly depending on the value of α as known by F.

Yet a third approach, which is the one to be adopted in this paper, is the "minimum sensitivity" method which is akin to the one used in control theory in feedback design. In our context (that is, the incentive design problem), L assumes a nominal value for the parameter α, say α^*, and designs a policy which renders the effect of variations in the value of α in a neighborhood of α^* on the L's objective functional, through F's optimal response function, a minimum. For this design method to be applicable, it is necessary that the nominal incentive problem (with $\alpha = \alpha^*$) admit more than one solution (which turns out to be the case if L has access to redundant information)—which generates an equivalence class of nominal solutions—so that a further selection could be made from the equivalence class of solutions thus generated to satisfy additional design criteria such as insensitivity. To be more explicit, if we denote the equivalence class of optimal policies for L, solving the nominal incentive design problem, by Γ_L^{eq}, and if α changes to a value $\alpha^* + \varepsilon$ from α^*, then F's response to a fixed $\gamma \in \Gamma_L^o$ will be $\beta_\gamma^{\alpha^*+\varepsilon}$, whose effect on J_L being

$$J_L(\gamma, \beta_\gamma^{\alpha^*+\varepsilon}) \triangleq \tilde{J}(\gamma,\varepsilon) \quad . \tag{11a}$$

Assuming that α is a scalar parameter, we formally expand $\tilde{J}$ into a Taylor series around $\varepsilon = 0$, to obtain

$$\tilde{J}(\gamma,\varepsilon) = \tilde{J}(\gamma,0) + \varepsilon\delta\tilde{J}(\gamma,0) + \frac{\varepsilon^2}{2!}\,\delta^2\tilde{J}(\gamma,0) + 0(\varepsilon^2) \tag{11b}$$

where the zero'th order term $\tilde{J}(\gamma,0) = J_L(\gamma,\beta_\gamma^{\alpha^*})$ is constant over Γ_L^{eq} and is the performance level desired by L, which corresponds to the response policy $\beta_\gamma^{\alpha^*} = v^d$ on part of F. Hence, the additional (higher order) terms in (11b) reflect the effect of the perturbation ε in the value of α on the performance of the leader. The first order perturbation (sensitivity) term $\delta\tilde{J}(\gamma,0)$ is generally either zero or is a constant over Γ_L^{eq}, and thus the leading term that determines the optimum choice out of Γ_L^{eq} is the second order sensitivity function $\delta^2\tilde{J}(\gamma,0)$ which is to be minimized over Γ_L^{eq} in order to desensitize L's performance to changes in the value of α, to second order. In certain cases, $\delta^2\tilde{J}(\gamma,0)$ is minimized not by a single element out of Γ_L^{eq} but by an entire (infinite) set of elements in Γ_L^{eq}. This then enables one to go to higher order terms in the expansion (11b) and minimize (desensitize) them using the available degrees of freedom. Viability of this approach will be demonstrated in the following sections in the context of a specific class of incentive decision problems.

3. Some Special Types of Incentive Decision Problems

In this section we identify some special types of incentive problems which have been considered before in the literature, and which fit the general framework introduced above.

A. *"Revelation of Truth"*

$$\eta_L(v,z_L) = v \ , \ z_F = \xi \ , \ \eta_F(\gamma,\xi) = (\gamma,\xi)$$

$$v^d(z_F) = \xi \ , \ J_L = \| v_\gamma^o - v^d \| \quad .$$

Here F knows the state of Nature which is not known by L. The question is whether it is possible for L to find a γ (whose argument is only v) for which $\| v_\gamma^o - v^d \| = 0 \Rightarrow v_\gamma^o = \xi$. In other words, such a policy will force F to "reveal the truth." A representative reference for this type of an incentive problem and its analysis is the book by Spence (1974).

B. *"Principal-Agent Problem"*

Let

$$\eta_L = \{z_L', z_L''\} \ , \ z_L' = \varphi(v,\xi) \ , \ z_L'' = z_L''(\xi)$$

where z_L' denotes the outcome of F's action v, which is observed by L (note that L does not observe v directly); furthermore, z_L'' is some other private information for L.

$$\eta_F = \{\gamma, z_F\} \ , \ z_F = z_F(\xi)$$

$$J_L = E\{L_L[\gamma(z_L', z_L'') - z_L']\}$$

$$J_F = E\ \{L_F[\gamma(z_L', z_L''),\ \beta(\eta_F)]\} \quad .$$

This is the so-called Principal-Agent Problem extensively studied in the economics literature (Shavel (1979), Mirrless (1976), Grossman and Hart (1983), Myerson (1983), Radner (1981) being an incomplete list of references—see also the survey articles by Jennergren (1980) and Ho, Luh and Olsder (1982)). The leader (L) is the principal, and F is the agent whose actions are not directly observed by L but only through the outcomes which also depend on ξ, the random state of Nature. L's policy γ is the fee schedule he should adopt so that his total average cost J_L (which depends on the difference between what he pays to the agent and what he gets in return) is minimized. F's optimal response, on the other hand, is determined by his minimization of J_F which depends on the fee schedule and the level of his effort.

This problem could be posed as a Stackelberg or Nash game; however a more reasonable solution concept in this context turns out to be the *Pareto-optimal solution*:

$$\min_{\gamma} J_L \text{ such that } \min_{\beta} J_F \le J_F^*$$

where J_F^* is a level of cost (disutility) tolerable by F.

C. *"Strategic Information Transmission"*

Let

$$\gamma = (\gamma_1, \gamma_2) \ , \ \eta_L = z_L(\xi) \ , \ \eta_F = \eta_F(\gamma_2(z_L))$$

$$J_L = E\{L_F(\gamma_1(z_L)\ ,\ \beta(\eta_F)\ ,\ \xi)\}$$

$$J_L = E\{L_L(\gamma_1(z_L)\ ,\ \beta(\eta_F)\ ,\ \xi)\} \quad .$$

The interpretation of this incentive design problem which has recently been considered by Crawford and Sobel (1982) is the following: A better informed sender (which is L) sends a possibly noisy signal $[\gamma_2(z_L)]$ to a receiver (which is F) who then takes an action that determines the performance (welfare) of both. L's problem is to choose the best γ_2, and possibly γ_1 which helps only in obtaining a lower cost performance, so that taking into account the cost-minimizing (min J_F) behavior of F, leads to the lowest possible cost level for him. Note that if we have a team problem, i.e. $J_L \equiv J_F$, then the best signal to be sent to F would be the observation z_L itself; however if $J_L \neq J_F$, it may be more advantageous for L to send a distorted version of z_L to F, which makes the problem interesting and challenging.

D. *"Redundant Dynamic Information"*

Here let

$$\eta_L = \{z(\xi), y(\xi), v\} \quad , \eta_F = \{\gamma, z(\xi)\}$$

$$v = v^d \text{ desirable element in } \mathfrak{V}$$

$$J_F = J_F(\gamma, \beta, \alpha) \;;\; \alpha = \alpha^* \text{ (nominal parameter value)}$$

where z denotes the information common to both L and F, and y denotes L's private information regarding the value of ξ. Note that L's information here is redundant and dynamic, because he has access to both the realized value of F's action and the observation on which this action is based. This will be the problem we will treat in the remaining portion of this paper, with L's objective being to find a $\gamma = \gamma(\eta_L)$ so that the minimization problem

$$\min_{\beta} J_F(\gamma, \beta, \alpha) \tag{12}$$

leads to the desired action $v = v^d$ when $\alpha = \alpha^*$, and furthermore that this policy carries some insensitivity property as α varies around α^*. This could also be viewed as a model matching problem where the model to be matched is an optimization problem admitting the unique solution $v = v^d$, and the control variable which achieves this is the γ used in (12). What is needed is, if possible, a perfect matching for $\alpha = \alpha^*$, and some satisfactory near-matching for α in a neighborhood of α^*. This could be accomplished, as to be shown in the next section, by basically following the sensitivity approach discussed in Section 2.

4. Exact Model Matching for a Class of Incentive Problems

Because of space limitations, we are going to discuss here only the scalar case, corresponding to Problem D above, which however captures the essential features of the more general class of models. Results on higher dimensional models can be found in Başar (1984), Başar and Cansever (1984).

By a scalar model, we mean here that $U = V = \mathbb{R}$, and the two random variables z and y are scalar; ξ on the other hand could be a vector-valued random variable. Let us take $u^d(z,y)$, $v^d(z)$ as a desirable pair of random variables, as discussed earlier in Section 2, and let $L_F(u,v,\xi,\alpha)$ denote F's loss function parameterized by α which is a scalar.

The first question to settle is the existence and characterization of a (or a class of) $\gamma = \gamma(v,z,y)$ which would lead to perfect model matching at $\alpha = \alpha^*$. In other words,

$$\min_{\beta} J_F(\gamma,\beta,\alpha^*) \to v^d(z)$$

$$\gamma(v^d(z),z,y) = u^d(z,y) \quad ?$$

The following theorem says that this is "in general" possible.

<u>Theorem 4.1</u>: (Başar, 1984)

Let

$$\underset{\xi|y,z}{E} \{\frac{\partial}{\partial u} L_F(u^d(z,y),v^d(z),\xi,\alpha^*)\} \neq 0 \quad \text{a.e. } y,z \quad . \tag{13}$$

Then, perfect model matching is possible ((u^d,v^d) is attainable) by a policy of the form

$$\gamma(v,z,y) = u^d(z,y) - Q(z,y)[v - v^d(z)] \tag{14}$$

where Q is any solution of

$$\underset{y|z}{E} \{Q(z,y)\ F(z,y)\} = G(z) \tag{15}$$

with

$$F(z,y) \triangleq \underset{\xi|y,z}{E} \{\frac{\partial}{\partial u} L_F(u^d(z,y),v^d(z),\xi,\alpha^*)\} \tag{16a}$$

$$G(z) \triangleq \underset{\xi,y|z}{E} \{\frac{\partial}{\partial v} L_F(u^d(z,y),v^d(z),\xi,\alpha^*)\} \tag{16b}$$

□

An important observation that can be made at this point is that the (z,y)-measurable random variable Q satisfying (15) is in general not unique, thus leading to an equivalence class of linear-in-v policies for L which make perfect matching possible. A clear demonstration of this is the construction of a class of solutions to (15):

$$Q(z,y) = g(z,y)G(z)/ \underset{y|z}{E} \{g(z,y)\ F(z,y)\} \tag{17}$$

where g is any (z,y)-measurable random variable satisfying the condition

$$\underset{y|z}{E} \{g(z,y)\ F(z,y)\} \neq 0 \ , \ \text{a.e. } z \quad . \tag{18}$$

Even this class (characterized by g) is sufficiently rich and has in general infinitely many elements. The corresponding class of policies (14) is an infinite subset of Γ_L^{eq} introduced in Section 2; let us denote this class by $\tilde{\Gamma}_L^{eq}$. Now, as in Section 2, this richness in the class of policies that makes perfect matching possible prompts yet another design goal which is to make F's response be minimally sensitive to changes in the value of α. Following the outlined sensitivity analysis of Section 2, we first compute the first order sensitivity function by taking L's cost function as

$$J_L = \|\beta_\gamma^\alpha - v^d\| \quad . \tag{19}$$

Using the notation of (11), we have

$$\delta\tilde{J}(\gamma,0) = d\beta_\gamma^\alpha(z)/d\alpha\Big|_{\alpha=\alpha^*} \underset{\xi,y|z}{E} \{\frac{d^2}{dud\alpha} L_F \cdot Q(z,y) - \frac{d^2}{dvd\alpha} L_F\}$$
$$/[\underset{\xi,y|z}{E} \{\frac{d^2}{du^2} L \cdot Q^2 - 2\frac{d^2}{dudv} L \cdot Q + \frac{d^2}{dv^2} L\}] \tag{20}$$

where the denominator is nonzero (in fact, positive) if $L_F(u,v,\xi,\alpha)$ is strictly convex in the pair (u,v). This expression can be made zero by choosing g to satisfy

$$\underset{y|z}{E} \{g(z,y)\, f_1(z,y)\} = 0 \tag{21a}$$

where

$$f_1(z,y) \triangleq F(z,y) \underset{\xi,y|z}{E} \{\frac{d^2}{dvd\alpha} L_F(u^d,v^d,\xi,\alpha^*)\} - G(z) \underset{\xi|y,z}{E} \{\frac{d^2}{dud\alpha} L_F(u^d,v^d,\xi,\alpha^*)\} . \tag{21b}$$

Hence, $\beta_\gamma^{\alpha^*+\varepsilon}$ can be made insensitive to first order in ε, over $\tilde{\Gamma}_L^{eq}$, by choosing γ as given by (14), with Q given by (17), and g(z,y) satisfying (18) and (21a). Such a policy indeed exists (generically), with one class of candidates for g(z,y) being

$$g(z,y) = \underset{y|z}{E} \{g_1(z,y)\, f_1(z,y)\} - g_1(z,y) \underset{y|z}{E} \{f_1(z,y)\} \tag{22}$$

where $g_1(z,y)$ is an arbitrary (z,y)-measurable random variable. The fact that (22) is a solution to (21a) can be checked by inspection, and condition (18) is satisfied provided that for every z-measurable random variable k(z),

$$\underset{\xi|y,z}{E} \{\frac{d^2}{dud\alpha} L_F(u^d,v^d,\xi,\alpha^*)\} \neq k(z) \underset{\xi|y,z}{E} \{\frac{d}{du} L_F(u^d,v^d,\xi,\alpha^*)\} \quad . \tag{23}$$

This then leads us to the following Theorem:

Theorem 4.2:

In addition to the hypothesis of Theorem 4.1, assume that (23) holds. Then, for the scalar version of the stochastic incentive design problem "D", perfect matching at a nominal value α^* of α, and insensitivity to first order (when α lies in a neighborhood of α^*) is achieved by a policy of the form

$$\gamma(v,z,y) = u^d(z,y) - Q(z,y)[v - v^d(z)]$$

where Q(z,y) is given by (17) and g(z,y) by (22) with $g_1(z,y)$ being an arbitrary (z,y)-measurable random variable. □

Now, since we have arbitrariness in the choice of g_1, we could go ahead and seek to annihilate the second-order sensitivity function with the additional degrees of freedom at our disposal. Calculating the second-order sensitivity function, we obtain

$$\delta^2\tilde{J}(\gamma,0) = d^2\beta_\gamma^\alpha(z)/d\alpha\Big|_{\alpha=\alpha^*} = \underset{\xi,y|z}{E} \{\frac{d^3}{dud^2\alpha} L_F \cdot Q - \frac{d^3}{dvd\alpha^2} L_F\} /$$
$$\cdot [\underset{\xi,y|z}{E} \{\frac{d^2}{du^2} L_F \cdot Q^2 - 2\frac{d^2}{dudv} L_F \cdot Q + \frac{d^2}{dv^2} L_F\}]$$

where the denominator is again positive because of strict convexity of L_F in (u,v).

The numerator can be made zero by choosing $g(z,y)$ to satisfy

$$\underset{y|z}{E}\ \{g(z,y)\ f_2(z,y)\} = 0 \tag{25a}$$

where

$$f_2(z,y) \triangleq F(z,y) \underset{\xi,y|z}{E} \{\frac{d^3}{dvd\alpha^2} L_F(u^d,v^d,\xi,\alpha^*)\} - G(z) \underset{\xi|y,z}{E} \{\frac{d^3}{dud\alpha^2} L_F(u^d,v^d,\xi,\alpha^*)\} . \tag{25b}$$

Hence, for also the second-order sensitivity function to be zero, we require $g(z,y)$ to be orthogonal to $f_2(z,y)$, in addition to being orthogonal to $f_1(z,y)$, under the conditional measure $\mathbb{P}(y|z)$, which is generically possible because $g(z,y)$ belongs to an (infinite dimensional) Hilbert space of random variables. Condition (18) on the other hand, translates to (as a counterpart of (23)): For every z-measurable random variable $k(z)$,

$$\underset{\xi|y,z}{E} \{\frac{d^3}{dud\alpha^2} L_F(u^d,v^d,\xi,\alpha^*)\} \neq k(z) \underset{\xi|y,z}{E} \{\frac{d}{du} L_F(u^d,v^d,\xi,\alpha^*)\} \quad . \tag{26}$$

This then leads to our third theorem:

Theorem 4.3:

In addition to the hypotheses of Theorems 4.1 and 4.2, let condition (26) hold. Then, perfect matching at the nominal value $\alpha = \alpha^*$, and insensitivity to second-order is possible by a policy of the form (14), where $Q(z,y)$ is given by (17) and $g(z,y)$ satisfies both (21a) and (25a). □

This procedure can be followed up to higher orders, and since $g(z,y)$ belongs to an infinite-dimensional Hilbert space it can be made orthogonal to a countable number of linearly independent random variables; hence, generically, sensitivity functions of arbitrary order could be annihilated by an appropriate linear-in-v policy of L. Details of this can be found in Başar and Cansever (1984).

We conclude this paper with a numerical illustration of the foregoing results in the next section.

5. An Illustrative Example

Let $\xi = (x,w_1,w_2)$ be a Gaussian random vector with mean zero and covariance the 3-dimensional identity matrix. Let

$z = x + w_1$ (common measurement)

$y = x + w_2$ (private measurement to L)

$u^d(z,y) = \frac{z}{2} + \frac{y}{3} + \frac{1}{3}$

$v^d(z) = \frac{z}{2} + \frac{1}{3}$

and F's loss function be

$$L_F(u,v,\xi,\alpha) = (x + u + v - \alpha)^2 + v^2$$

with $\alpha^* = 2$. Note that L_F is strictly convex in the pair (u,v).

Here condition (13) is satisfied, and solution (17) to equation (15) can be written as

$$Q(z,y) = \frac{(13z-2)g(z,y)}{2 \underset{y|z}{E} \{g(z,y)(2y+4z-3)\}} \tag{27}$$

under the condition that

$$\underset{y|z}{E} \{g(z,y)(2y+4z-3)\} \neq 0 \quad . \tag{28}$$

Equation (21a), whose solution annihilates the first-order sensitivity function, can be rewritten as

$$\underset{y|z}{E} \{g(z,y)[17z+2y-5]\} = 0 \quad .$$

A possible solution to this which also satisfies condition (28) is

$$g(z,y) = -3z^2 + 6zy + 4y - 5z + 10$$

with the corresponding $Q(z,y)$ being

$$Q^0(z,y) = (-3z^2 + 6zy + 4y - 5z + 10)/12 \quad . \tag{29}$$

This solution has the property that it annihilates not only the first-order sensitivity function but also sensitivity functions of higher (arbitrary) order. This can best be seen by minimizing

$$\underset{\xi,y|z}{E} \{L_F(u^d + Q^0(v - v^d), v, \xi, \alpha)\}$$

over $v \in \mathbb{R}$ for each fixed $z \in \mathbb{R}$ and for arbitrary α. The unique solution is

$$\beta_\gamma^\alpha(z) = \frac{z}{2} + \frac{1}{3} \equiv v^d(z)$$

which is the desired one independently of the precise value of α. This is also seen in the plot of Figure 1 where the dashed line represents F's optimum reaction when $z = -0.5$ and α is arbitrary. The solid line, on the other hand, represents F's optimum reaction to a best policy of L obtained under the restriction of only z-measurable gain coefficient Q, which is

$$Q(z) = (13z - 2)/(10z - 4) \quad . \tag{30}$$

Note that in this case F's response varies with α.

6. Acknowledgement

Some of the results reported herein were obtained jointly with D. H. Cansever, and research was supported in part by the Office of Naval Research under Contract N00014-82-K-0469, in part by the Air Force Office of Scientific Research under Contract AFOSR-84-0054, and in part by the U. S. Department of Energy, Electric Energy Systems Division, under Contract DE-AC01-81RA-50658, with Dynamic Systems, P. O. Box 423, Urbana, Illinois 61801.

7. References

Başar, T. (1984), "Affine Incentive Schemes for Stochastic Systems with Dynamic Information," SIAM J. Control and Optimization, vol. 22, no. 2, pp. 199-210.

Başar, T. and D. Cansever (1984), "Robust Incentive Policies for Stochastic Decision Problems in the Presence of Parametric Uncertainty," Proc. 9th IFAC World Congress, Budapest, Hungary, July 2-6.

Başar, T. and A. Haurie (1984), "Feedback Equilibria in Differential Games with Structural and Modal Uncertainties," in Advances in Large Scale Systems, vol. 1, J. B. Cruz, Jr. (Ed.), JAI Press, Inc., Connecticut.

Crawford, V. P. and J. Sobel (1982), "Strategic Information Transmission," Econometrica, vol. 50, no. 6, pp. 1431-1451.

Grossman, S. J. and O. D. Hart (1983), "On the Analysis of the Principal-Agent Problem," Econometrica, vol. 51, no. 1, pp. 7-45.

Harsanyi, J. C. (1968), "Games with Incomplete Information played by "Bayesian" Players. Parts I, II and III," Management Science, vol. 14, nos. 3, 5, 7, pp. 153-182, 320-334, 486-502.

Ho, Y. C., P. Luh and G. J. Olsder (1982), "A Control Theoretic View on Incentives," Automatica, vol. 18, pp. 167-179.

Jennergren, L. P. (1980), "On the Design of Incentives in Business Firms — A Survey of Some Recent Results," Management Science, vol. 26, pp. 180-201.

Mirrless, J. A. (1976), "The Optimal Structure of Incentives and Authority Within an Organization," Bell Journal of Economics, vol. 7, pp. 105-131.

Myerson, K. B. (1983), "Mechanism Design by an Informed Principal," Econometrica, vol. 51, no. 6, pp. 1767-1798.

Radner, R. (1981), "Monitoring Cooperative Agreements in a Repeated Principal-Agent Relationship," Econometrica, vol. 49, no. 5, pp. 1127-1148.

Shavel, S. (1979), "Risk Sharing and Incentives in the Principal and Agent Relationship," Bell J. Economics, vol. 10, pp. 55-73.

Spence, M. (1974), Market Signaling, Cambridge, MA: Harvard Univ. Press.

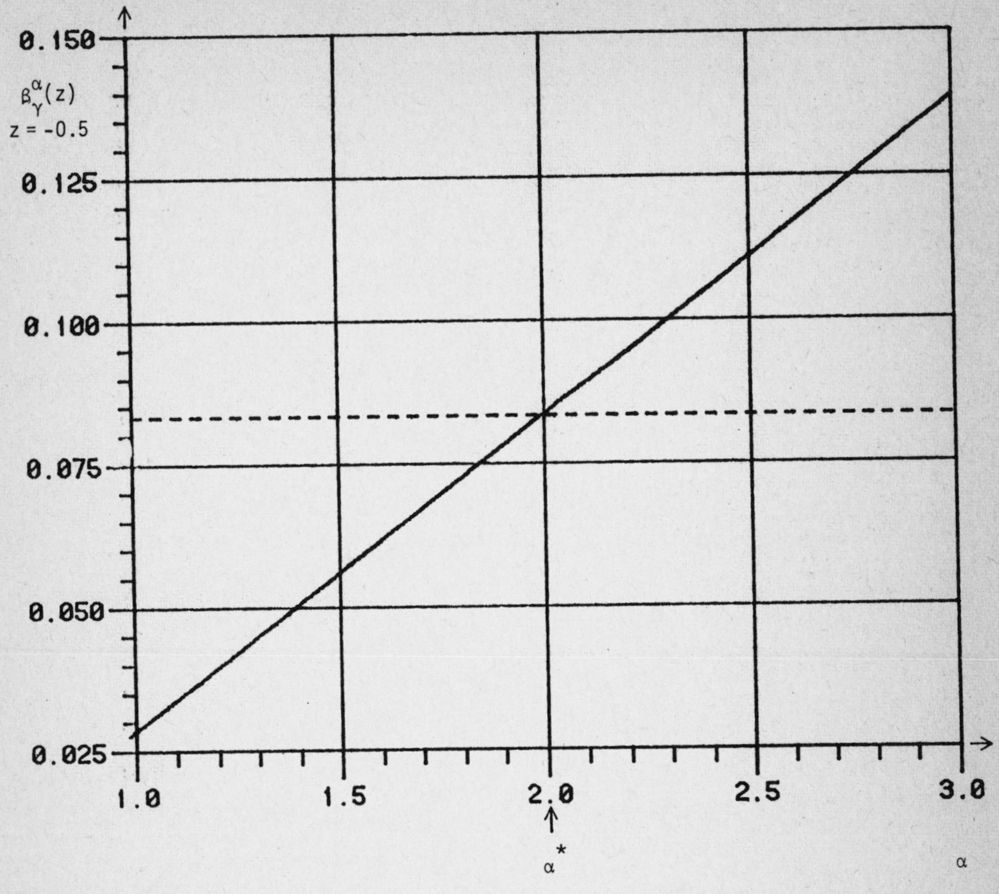

Figure 1: In the example of Section 5, optimum response of the follower to the leader's optimal policy (dashed line) and to L's optimal policy which uses z-measurable gain (30) (solid line) for different values of α. [Here $z = -0.5$.]

RECENT RESULTS ON NONLINEAR MODEL MATCHING

M.D. Di Benedetto and A. Isidori
Dipartimento di Informatica e Sistemistica
Università di Roma "La Sapienza"
Via Eudossiana, 18
00184 Roma

1. INTRODUCTION AND PROBLEM STATEMENT

The linear model matching problem, i.e. the problem of compensating a linear multivariable system in order to obtain the same transfer function as the one of a prescribed linear model, has been investigated and solved by several authors (see, e.g. [1],[2],[3]). In particular, a geometric approach was followed by Morse, who proposed a necessary and sufficient condition for the solvability of a linear model matching problem based on the construction of an appropriate controllability distribution.

Malabre [4] showed the equivalence between Morse's condition and a relation involving the "infinite zero structures" of the controlled system and of the model.

The problem of designing a compensating control for a nonlinear system in order to match a prescribed linear input-output behavior has been solved by Isidori [5]. The case in which the prescribed input-output behavior to match is that of a nonlinear model has been recently investigated by the authors [6].

In this paper, we summarize the main results of [6], i.e. the nonlinear version of the Morse's condition and its equivalent in terms of "infinite zero structures". Moreover, in the particular case of systems which are input-output linearizable via static state - feedback, we discuss the advantage of the condition based on the infinite zero structure.

In what follows, we shall consider a fixed nonlinear plant P, described by equations of the form

$$\dot{x} = f(x) + g(x)u \tag{1.1a}$$

$$y = h(x) \tag{1.1b}$$

with state $x \in X \subset \mathbb{R}^n$, input $u \in \mathbb{R}^n$ and output $y \in \mathbb{R}^p$. f and the m columns $g_1, g_2, \dots, g_m$ of the matrix g are real analytic vector fields on $\mathbb{R}^n$ and h is a real analytic function. We shall also assume that a model M is given, described by equations of the form

$$\dot{x}_M = f_M(x_M) + g_M(x_M)u_M \tag{1.2a}$$

$$y_M = h_M(x_M) \tag{1.2b}$$

with state $x_M \in X_M \subset \mathbb{R}^{n_M}$, input $u_M \in \mathbb{R}^{m_M}$, output $y_M \in \mathbb{R}^p$ and real analytic f_M, g_M, h_M.

In the present setting the compensator Q used to control the system (1.1) is described by equations of the form

$$\dot{z} = a(z,x) + b(z,x)u_M \tag{1.3a}$$

$$u = c(z,x) + d(z,x)u_M \tag{1.3b}$$

with state $z \in Z \subset \mathbb{R}^\nu$ and real analytic a,b,c,d. The compensated plant, denoted $P \circ Q$ (i.e. the composition of (1.1) and (1.3)), is clearly a dynamical system with the same structure as (1.1).

The goal of model matching synthesis is to find a,b,c,d such that the compensated plant displays the same input-output behavior as the model. To be more precise, let us recall that the output of a system like (1.1) may be given a Volterra series expansion, of the form

$$y(t) = w_o(t,x^o) + \sum_{i=1}^{m} \int_0^t w_i(t,\tau_1,x^o)u_i(\tau_1)d\tau_1 + $$
$$+ \sum_{i_1,i_2=1}^{m} \int_0^t \int_0^{\tau_1} w_{i_1 i_2}(t,\tau_1,\tau_2,x^o)u_{i_1}(\tau_1)u_{i_2}(\tau_2)d\tau_1 d\tau_2 + \dots$$

where x^o is the initial state at time $t = 0$.

In what follows, we require the compensator Q to be able to allow the reproduction of any input-output behavior of the model from any initial state of the process. More precisely we seek a solution such that for every initial state of the process and every initial state of the model there exists an initial state of the compensator for which all the Volterra kernels of the model (denoted w^M) and of the compensated plant (denoted $w^{P \circ Q}$) coincide.

The above considerations lead to consider the following formulation.

Nonlinear Model Matching Problem (MMP). Given a plant $P = (f,g,h)$, a model $M = (f_M, g_M, h_M)$ and a point $(x, x_M) \in \mathbb{R}^n \times \mathbb{R}^{n_M}$, find neighborhoods U of x and U_M of x_M, an integer ν, an open subset V of $\mathbb{R}^\nu$, a quadruple (a,b,c,d) with a,b,c,d analytic functions defined on $V \times U$ and a map $F : U \times U_M \to V$ such that

$$w^{P\circ Q}_{j_1 \dots j_i}(t,\tau_1,\dots,\tau_i,x,F(x,x_M)) = w^{M}_{j_1 \dots j_i}(t,\tau_1,\dots,\tau_i,x_M)$$

for all $i \geq 1$, for all $1 \leq j_i \leq m_M$ and for all (x,x_M) in $U \times U_M$.

Note that the "drift terms" are not required to be the same. Moreover, local solutions are sought, since global ones might be difficult to find.

2. THE GEOMETRIC APPROACH TO NONLINEAR MODEL MATCHING: A REVIEW

In this section, some recent results presented in [6] are reviewed. It is shown that the possibility of solving an MMP may be expressed either in terms of properties of a suitable controlled invariant distribution or in terms of the so-called "infinite zero structures" associated with the process and the model.

It is assumed, in what follows, that the reader is familiar with basic concepts and notations used in the differential - geometric approach to nonlinear control theory (see e.g. [7]). For the sake of completeness, let us just recall here some definitions concerning invariant distributions.

Given a control system of the form (1.1), defined on a manifold X, a distribution Δ on X is *controlled invariant* if there exist feedback functions α and β such that

$$[f + g\alpha,\ \Delta] \subset \Delta \qquad (2.1a)$$

$$[g\beta,\ \Delta] \subset \Delta \qquad (2.1b)$$

If β is a nonsingular matrix, then (2.1) imply

$$[f,\ \Delta] \subset \Delta + \mathrm{span}\{g\} \qquad (2.2a)$$

$$[g,\ \Delta] \subset \Delta + \mathrm{span}\{g\} \qquad (2.2b)$$

Conversely, if Δ is involutive and the distributions Δ, span{g} and Δ + span{g} have constant dimension, then (2.2) imply the existence of local feedback functions such that (2.1) are satisfied.

In what follows, $\mathbb{I}(f,g,K)$ will denote the class of distributions which satisfy (2.2) and are contained in a given distribution K on X.

The first result of this section provides a sufficient condition for the existence of a solution of an MMP. The statement of this condition requires the introduction of some additional notations. With a given MMP, i.e. a process $P = (f,g,h)$ and a model $M = (f_M, g_M, h_M)$, we may associate the vector fields

$$\hat{f}(\hat{x}) = \begin{pmatrix} f(x) \\ \\ f_M(x_M) \end{pmatrix}, \quad \hat{g}(\hat{x}) = \begin{pmatrix} g(x) \\ \\ 0 \end{pmatrix}, \quad \hat{p}(\hat{x}) = \begin{pmatrix} 0 \\ \\ g_M(x_M) \end{pmatrix}$$

and the function

$$\hat{h}(\hat{x}) = h(x) - h_M(x_M)$$

(with $\hat{x} = (x, x_M)$), which are uniquely defined.

We are now ready to state the following

THEOREM 2.1. Let $\hat{\Delta}^*$ be the unique maximal element of $\mathbb{I}(\hat{f},\hat{g},(d\hat{h})^{\perp})$. Assume $\hat{\Delta}^*$, span$\{\hat{g}\}$and $\hat{\Delta}^*$ + span$\{\hat{g}\}$ have constant dimension. If

$$\text{span}\{\hat{p}\} \subset \hat{\Delta}^* + \text{span}\{\hat{g}\} \tag{2.3}$$

the MMP is solvable.

PROOF. The distribution $\hat{\Delta}^*$ is locally controlled invariant. Therefore, around every point in $X \times X_M$ it is possible to find a function $\hat{\alpha}$ which makes the condition

$$[\hat{f} + \hat{g}\hat{\alpha}, \hat{\Delta}^*] \subset \hat{\Delta}^*$$

satisfied. Moreover, from (2.3), we deduce that around every point in $X \times X_M$ it is possible to find a function $\hat{\gamma}$ such that

$$\text{span}\{\hat{g}\hat{\gamma} + \hat{p}\} \subset \hat{\Delta}^*$$

The above inequalities, since $\hat{\Delta}^* \subset (d\hat{h})^{\perp}$ by definition, are readily seen to imply

$$L_{(\hat{g}\hat{\gamma}+\hat{p})} L^k_{(\hat{f}+\hat{g}\hat{\alpha})} \hat{h}(\hat{x}) = 0$$

for all $k \geq 0$, for all $\hat{x}$ in the neighborhood where $\hat{\alpha}$ and $\hat{\gamma}$ are both defined.

This, in turn, implies that in the system

$$\dot{x} = f(x) + g(x)\hat{\alpha}(x,x_M) + g(x)\hat{\gamma}(x,x_M)u_M$$

$$\dot{x}_M = f_M(x_M) + g_M(x_M)u_M$$

$$\hat{y} = h(x) - h_M(x_M)$$

the output $\hat{y}$ is independent of the input u_M for any possible initial state $\hat{x}^o = (x^o, x_M^o)$.

Now, observe that the output $\hat{y}$ of this system may be interpreted as the difference between the output of the system

$$\dot{x} = f(x) + g(x)\hat{\alpha}(x,x_M) + g(x)\hat{\gamma}(x,x_M)u_M$$

$$\dot{x}_M = f_M(x_M) + g_M(x_M)u_M \tag{2.5}$$

$$y = h(x)$$

initialized at (x^o, x_M^o), and the one of the model, initialized at x_M^o. Since this difference is independent of u_M, the output of (2.5) and the one of the model differ only by a term which doesn't depend on u_M, i.e. a "drift term".

Since the system (2.5) may be viewed as the process P, initialized in x^o, composed with a compensator Q defined by the equations

$$\begin{aligned} \dot{z} &= f_M(z) + g_M(z)u_M \\ u &= \hat{\alpha}(x,z) + \hat{\gamma}(x,z)u_M \end{aligned} \tag{2.6}$$

initialized at $z^o = x_M^o$, the previous conclusion shows that the compensator (2.6) solves the MMP. □

When dealing with some particular classes of systems, e.g. linear systems, the condition (2.3) is also necessary for the solvability of an MMP. In the general case of nonlinear systems, the proof of the necessity of (2.3) requires some additional assumptions, essentially because the relation (2.1a) alone does not imply both relations (2.2).

In order to introduce these extra assumptions, it is useful to remark that the difference between the output of the compensator plant $P \circ Q$ and that of the model M may be viewed as the output of an extended system

$$\begin{aligned} \dot{\bar{x}} &= \bar{f}(\bar{x}) + \bar{g}(\bar{x})\bar{u} \\ \bar{y} &= \bar{h}(\bar{x}) \end{aligned} \tag{2.7}$$

with

$$\bar{x} = (x, x_M, z)$$

$$\bar{f}(\bar{x}) = \begin{pmatrix} f(x) \\ f_M(x_M) \\ 0 \end{pmatrix}, \quad \bar{g}(\bar{x}) = \begin{pmatrix} g(x) & 0 & 0 \\ 0 & g_M(x_M) & 0 \\ 0 & 0 & I \end{pmatrix}$$

$$\bar{h}(\bar{x}) = h(x) - h_M(x_M)$$

subject to a feedback control law of the form

$$\bar{u} = \bar{\alpha}(\bar{x}) + \bar{\beta}(\bar{x})u_M \tag{2.8}$$

with

$$\bar{\alpha}(\bar{x}) = \begin{pmatrix} c(z,x) \\ 0 \\ a(z,x) \end{pmatrix}, \quad \bar{\beta}(\bar{x}) = \begin{pmatrix} d(z,x) \\ I \\ b(z,x) \end{pmatrix}$$

If the compensator Q solves the MMP, the Volterra kernels of (2.7) composed with (2.8) vanish in the initial state $\bar{x}^o = (x^o, x_M^o, F(x^o, x_M^o))$. This, in particular, implies

$$L_{(\bar{g}\bar{\beta})} L^k_{(\bar{f}+\bar{g}\bar{\alpha})} \bar{h}(\bar{x}^o) = 0$$

for all $k \geq 0$. From this we may deduce that the largest distribution, denoted $\bar{\Delta}_{max}$, invariant under $(\bar{f}+\bar{g}\bar{\alpha})$ and contained in $(d\bar{h})^{\perp}$ satisfies the condition

$$\text{span}\{\bar{g}\bar{\beta}\}(\bar{x}) \subset \bar{\Delta}_{max}(\bar{x}) \tag{2.9}$$

at all $\bar{x}$ in the subset M of $U \times U_M \times V$

$$M = \{(x, x_M, z) \in U \times U_M \times V \mid z = F(x, x_M)\}$$

If one assumes that $\bar{\Delta}_{max}$ is an element of $\mathbb{I}(\bar{f}, \bar{g}, (d\bar{h})^{\perp})$ or, at least, is contained in the maximal element $\bar{\Delta}^*$ of $\mathbb{I}(\bar{f}, \bar{g}, (d\bar{h})^{\perp})$ then, the projection of

$$\text{span}\{\bar{g}\bar{\beta}\}(\bar{x}) \subset \bar{\Delta}^*(\bar{x})$$

onto $U \times U_M$ yields the desired condition (2.3).

This is summarized in the following statement, in which we use $\hat{\hat{g}}$ to denote the set of vector fields

$$\hat{\hat{g}}(x,x_M) = \begin{pmatrix} g(x) & 0 \\ 0 & g_M(x_M) \end{pmatrix}$$

THEOREM 2.2. Suppose the MMP is solved by some controller Q. Let $\bar{\Delta}_{max}$ be the largest distribution invariant under $(\bar{f}+\bar{g}\bar{\alpha})$ and contained in $(d\bar{h})^{\perp}$ and $\bar{\Delta}^*$ the unique maximal element of $\mathbb{I}(\bar{f},\bar{g},(d\bar{h})^{\perp})$. Assume $\bar{\Delta}^*$ is nonsingular and $\bar{\Delta}_{max}$ is such that

$$\bar{\Delta}_{max} \subset \bar{\Delta}^*$$

Then

$$\text{span}\{\hat{p}\} \subset \hat{\hat{\Delta}}^* + \text{span}\{\hat{g}\} \tag{2.10}$$

where $\hat{\hat{\Delta}}^*$ is the unique maximal element of $\mathbb{I}(\hat{f},\hat{\hat{g}},(d\hat{h})^{\perp})$. Moreover, condition (2.10) implies condition (2.3).

Details on the proof of this Theorem may be found in [6].

In a recent paper [4], Malabre has shown that, in the case of linear systems, the existence of a solution of an MMP depends on a relation which involves the "infinite zero structures" of the system and of the model. His proof consists in showing the equivalence between that relation and the linear version of (2.10). Essentially the same equivalence may be shown to hold in the present nonlinear setting. To this end, let us recall that the infinite zero structure of a triple (f,g,h) may be defined, following Nijmeijer and Schumacher [8], as the sequence of integers

$$p^1 = \dim(G^{\perp} + \Omega^*) - \dim(G^{\perp})$$

$$p^k = \dim(G^{\perp} + \Omega^*) - \dim(G^{\perp} + \Omega_{k-2}) \qquad k \geq 2$$

where $G = \text{span}\{g\}$ and the sequence of codistributions $\Omega_o, \Omega_1, \ldots, \Omega^*$ is generated by the algorithm

$$\Omega_o = dh$$

$$\Omega_k = \Omega_{k-1} + \sum_{i=0}^{m} L_{g_i}(G^{\perp} \cap \Omega_{k-1}) \tag{2.11}$$

where $g_o := f$.

In the same way, we can associate a list of integers $\hat{p}^k$ to the triple $(\hat{f},\hat{\hat{g}},\hat{h})$. Obviously, these definitions make sense under the assumption that all the codistributions involved have constant dimension.

The following result, whose proof may again be found in [6], provides an alternative way of checking the condition (2.10).

THEOREM 2.3. Suppose the sequences p^k and $\hat{p}^k$, $k \geq 1$, are defined. The condition (2.10) is satisfied if and only if

$$p^k = \hat{p}^k \qquad \text{for all } k \geq 1 \tag{2.12}$$

3. THE CASE OF LINEARIZABLE SYSTEMS

In this section, we consider the case where the given model M is linear and the plant P can be made linear, via static state-feedback, from an input-output point of view. In this particular case, the condition (2.12) has been proved to be necessary and sufficient for the existence of a solution of an MMP (see [5]). The aim of this section is to show that, in the present case of a linear model M and a linearizable plant P, testing the condition (2.12), based on the evaluation of the infinite zero structure of appropriate systems, is easier than testing the equivalent condition (2.10), based on the construction of a suitable controlled invariant distribution.

In the general case, the computations involved in the test of either one of the conditions (2.10) and (2.12) are not substantially different, mainly because of the geometric characterization of the infinite zero structure for a nonlinear system. However, in the case being examined in this section, the infinite zero structure may be computed directly on parameters which characterize the input-output behavior of the system.

The particular class of systems we are dealing with has been introduced in [5]. The main conclusion of this investigation was the proof of the equivalence between the following facts:

(i) there exists an invertible static state feedback, i.e. a control law of the form

$$u = \alpha(x) + \beta(x)v$$

with invertible β, under which the input-output behavior of system (1.1) takes the form

$$y(t) = w_o(t,x^o) + \sum_{i=1}^{m}\int_o^t w_i(t-\tau)u_i(\tau)d\tau$$

(ii) the formal power series $T(s,x)$, uniquely associated with the process $P = (f,g,h)$ and defined as

$$T(s,x) = \sum_{k=0}^{\infty} T_k(x) s^{-(k+1)}$$

where

$$T_k(x) = L_g L_f^k h(x)$$

may be factored in the form

$$T(s,x) = K(s) R(s,x) \tag{3.1}$$

where $K(s)$ is a formal power series whose coefficients are $p \times m$ matrices of real numbers

$$K(s) = \sum_{k=0}^{\infty} K_k s^{-(k+1)}$$

and $R(s,x)$ is an invertible formal power series whose coefficients are $m \times m$ matrices of real analytic functions

$$R(s,x) = R(x) + \sum_{k=0}^{\infty} R_k(x) s^{-(k+1)}$$

(iii) the sequence of Toeplitz matrices

$$\Theta_k(x) = \begin{pmatrix} T_o(x) & T_1(x) & \dots & T_k(x) \\ 0 & T_o(x) & \dots & T_{k-1}(x) \\ \dots\dots & & & \\ 0 & 0 & \dots & T_o(x) \end{pmatrix} \tag{3.2}$$

is such that

$$\rho_{\mathbb{R}}(\Theta_k) = \rho_{\mathbf{M}}(\Theta_k)$$

for all $k \geq 0$, where $\rho_{\mathbb{R}}(\Theta)$ denotes the dimension of the $\mathbb{R}$-vector space generated by the rows of Θ and $\rho_{\mathbf{M}}(\Theta)$ denotes the dimension of the $\mathbf{M}$-vector space generated by the rows of Θ, $\mathbf{M}$ being the field of meromorphic functions.

The existence of the factorization (3.1) clearly enables us to associate with any system of this particular class an "infinite zero structure" by simply taking the infinite zero structure of $K(s)$. The latter is defined via the so-called Smith-McMillan factorization at infinity

$$K(s) = L(s)\Lambda(s)D(s)$$

in which L(s) and D(s) are bicausal (i.e. invertible) formal power series and $\Lambda(s)$ takes the form

$$\Lambda(s) = \mathrm{diag}\{I_{\delta_1}\frac{1}{s},\ldots,I_{\delta_q}\frac{1}{s^q},0\}$$

The string of integers $\delta_1,\ldots,\delta_q$, which is uniquely associated with K(s) and characterizes its infinite zero structure, is also uniquely related to the infinite zero structure of the nonlinear system P = (f,g,h), as a consequence of the following result.

LEMMA 3.1. If the system (1.1) is such that any one of the three equivalent conditions (i),(ii) or (iii) is satisfied, then

$$p^k = \sum_{i=k}^{q} \delta_i \tag{3.3}$$

On the other hand, the computation of the string $\delta_1,\ldots,\delta_q$, unlike the one of the related string $p^1,\ldots,p^q$, does not involve the construction of suitable codistributions (see algorithm (2.11)) but may be entirely carried out in terms of rank evaluation of the Toeplitz matrices (3.2). As a matter of fact, as in [9] one may easily see that

$$\rho_{\mathbb{R}}(\Theta_k) = (k+1)\delta_1 + k\delta_2 + \ldots + \delta_{k+1} \tag{3.4}$$

Thus, the combined use of (3.3) and (3.4) provides an alternative and - possibly - easier way to evaluate the infinite zero structure of P.

If the model to be matched is also linear, then the triple $(\hat{f},\hat{g},\hat{h})$ clearly belongs to the same class of systems and its infinite zero structure may still be computed via (3.3) and (3.4).

4. CONCLUSIONS

The purpose of this paper was a review of some recent results dealing with the problem of matching a prescribed nonlinear input-output behavior via dynamic state feedback. It has been shown that, under appropriate hypotheses, the solvability of a model matching problem can be expressed in terms of properties of a suitable controlled invariant distribution. Those properties have been related to the infinite zero structures of the process and the model. Then, the case in which the plant can be made linear, from an input-output point of view, via a static state feedback, and the model is linear has been discussed. In that case, it has been shown that the infinite zero structure can be computed directly on the coefficients of the Taylor-series expansion of the first-order Volterra kernels. This simplifies the test for the existence of a solution of the model matching problem.

REFERENCES

[1] B.C. MOORE, L.M. SILVERMAN: "Model Matching by State Feedback and Dynamic Compensation". IEEE Trans. Automat. Contr., Vol.AC-17 (1972), 491-497.

[2] L.M. SILVERMAN: "Inversion of Multivariable Linear Systems". IEEE Trans. Automat. Contr., Vol. AC-14 (1969), 270-276.

[3] A.S. MORSE: "Structure and Design of Linear Model Following Systems". IEEE Trans. Automat. Contr., Vol. AC-18 (1973), 346-354.

[4] M. MALABRE: "Structure à l'infini des triplets invariants; application à la poursuite parfaite de modèle". 5-th Int. Conf. on Analysis and Optimization of Systems, I.N.R.I.A., Paris, (1982), 43-53.

[5] A. ISIDORI: "The Matching of a Prescribed Linear Input-Output Behavior in a Nonlinear System". To appear on IEEE Trans. Automat. Contr., Vol. AC-30, (1985).

[6] M.D. DI BENEDETTO, A. ISIDORI: "The Matching of Nonlinear Models via Dynamic State Feedback". Report 04.84, Dip. Informatica e Sistemistica, Università di Roma "La Sapienza", (1984).

[7] A. ISIDORI, A.J. KRENER, C. GORI-GIORGI, S. MONACO: "Nonlinear Decoupling via Feedback: A Differential Geometric Approach". IEEE Trans. Automat. Contr., Vol. AC-26 (1981), 331-345.

[8] H. NIJMEIJER , J.M. SCHUMACHER: "Zeros at infinity for affine nonlinear control systems". Memorandum n. 441, Twente University of Technology, (1983).

[9] P.M. VAN DOOREN, P. DEWILDE, J. WANDEWALLE: "On the determination of the Smith-MacMillan form of a Rational Matrix from its Laurent expansion". IEEE Trans. Circuits and Systems, Vol. CAS-26, (1979), 180-189.

A CRITERION FOR OPTIMALITY AND ITS APPLICATION TO PARAMETRIC PROGRAMMING

Bruno Brosowski
J.W.Goethe-Universität Frankfurt
Fachbereich Mathematik
Robert-Mayer-Str. 6 - 10
D-6000 Frankfurt/West Germany

INTRODUCTION

This paper deals with an optimality criterion which was derived earlier by KRABS [4,p.163]. We give conditions for the validity of this criterion and apply it to the investigation of the sovability set in the case of a variable restriction vector. This result lead to a characterization of a set of optimal points and a condition for the upper semicontinuity of the optimal set mapping.

I. THE MINIMIZATION PROBLEM

Let T be a compact Hausdorff-space, U be a non-empty open subset of $\mathbb{R}^N$. For each triplet

$$\sigma := (A,b,p)$$

of continuous mappings

$$A : T \times U \longrightarrow \mathbb{R}, \quad b : T \longrightarrow \mathbb{R}, \quad p : U \longrightarrow \mathbb{R}$$

consider the minimization problem

MP(σ). Minimize $p : U \longrightarrow \mathbb{R}$ subject to the side conditions

$$\forall_{t\in T} \; A(t,u) \leq b(t).$$

We explain this setting by some examples.

EXAMPLE 1.1. *Linear minimization.*

Let $U := \mathbb{R}^N$ and let $B : T \longrightarrow \mathbb{R}^N$ be a continuous mapping . Define a continuous mapping $A : T \times U \longrightarrow \mathbb{R}$ by setting

$$\forall_{t\in T} \; \forall_{u\in\mathbb{R}^N} \; A(t,u) := \langle B(t),u\rangle \;,$$

where $\langle\cdot,\cdot\rangle$ denotes the usual inner product. Let $\sigma' := (B,b,p)$, where $b \in C(T)$ and $p \in \mathbb{R}^N$. With these assumptions we receive the following minimization problem of type MP:

LMP(σ'). Minimize $\langle p,u\rangle$ subject to the side conditions

$$\forall_{t\in T} \; A(t,u) := \langle B(t),u\rangle \leq b(t),$$

i.e. we have the usual semi-infinite linear minimization problem.

EXAMPLE 1.2. *Generalized fractional minimization.*

Let $B : T \longrightarrow \mathbb{R}^N$ and $C : T \longrightarrow \mathbb{R}^N$ be continuous mappings such that the open convex subset

$$U_o := \bigcap_{t\in T} \{v \in \mathbb{R}^N \mid \langle C(t),v\rangle > 0\}$$

is non-empty. Let B_o, C_o be in $\mathbb{R}^N$, $\delta \geq 0$, and let $\gamma : T \longrightarrow \mathbb{R}$ be a continuous non-negative function. Define mappings

$$A : T \times U_o \times U \longrightarrow \mathbb{R} \quad \text{and} \quad p : U_o \times \mathbb{R} \longrightarrow \mathbb{R}$$

by setting

$$A(t,v,z) := \frac{\langle B(t),v\rangle}{\langle C(t),v\rangle} - \gamma(t)z$$

and

$$p(v,z) := \frac{\langle B_o,v\rangle}{\langle C_o,v\rangle} + \delta z.$$

For special choices of $B,C,B_o,C_o,\gamma,\delta$ and T one obtains the fractional minimization problem ($\delta = 0$, $\gamma = 0$), the rational Chebyshev approximation problem, compare BROSOWSKI and GUERREIRO [3].

Rational Chebyshev approximation.
Let

$$\{g_1,g_2,\ldots,g_n\} \quad \text{and} \quad \{h_1,h_2,\ldots,h_m\}$$

be real continuous functions defined on a compact Hausdorff space S and define the compact Hausdorff space $T := \{-1,1\} \times S$. For every $t = (\eta,s) \in T$, define the vectors

$$B(t) := \eta\,\overline{B}(s) := \eta(g_1(s),g_2(s),\ldots,g_n(s),0,0,\ldots,0),$$
$$C(t) := C(s) := (0,0,\ldots,0,h_1(s),h_2(s),\ldots,h_m(s))$$

of $\mathbb{R}^{n+m}$.

For every $(t,v,z) \in T \times U_o \times \mathbb{R}$ with

$$v = (\alpha_1,\alpha_2,\ldots,\alpha_n,\beta_1,\beta_2,\ldots,\beta_m),$$

we have

$$A(t,v,z) := \eta \frac{\langle\bar{B}(s),v\rangle}{\langle C(s),v\rangle} - \gamma(\eta,s)z$$

$$= \eta \frac{\sum_{i=1}^{n} \alpha_i g_i(s)}{\sum_{i=1}^{m} \beta_i h_i(s)} - \gamma(\eta,s)z.$$

Then for every $x \in C(S)$, the problem

<u>MPR(x).</u> Minimize $p(v,z) := z$
subject to

$$\forall_{(\eta,s)\in T} \quad \eta \frac{\sum_{i=1}^{n} \alpha_i g_i(s)}{\sum_{i=1}^{m} \beta_i h_i(s)} - \gamma(\eta,s)z \leq \eta x(s)$$

is a minimization problem of type MP(σ).

The problem MPR(x) is equivalent to certain rational Chebyshev approximation problems. In fact, consider

$$V := \left\{ \frac{\sum_{i=1}^{n} \alpha_i g_i}{\sum_{i=1}^{m} \beta_i h_i} \in C(S) \;\middle|\; \forall_{s\in S} \sum_{i=1}^{m} \beta_i h_i(s) > 0 \right\}.$$

If $\gamma(\eta,s) = w(s) > 0$, the problem MPR(x) is equivalent to the problem of finding a best rational Chebyshev approximation to x from V with weight function w, i.e. $(v_o,z_o) \in U_o \times \mathbb{R}$ with

$$v_o = (\alpha_{o1},\alpha_{o2},\ldots,\alpha_{on},\beta_{o1},\beta_{o2},\ldots,\beta_{om})$$

is a solution of MPR(x) iff

$$z_o = \left\| \frac{x - r_o}{w} \right\|_\infty = \inf_{v\in V} \left\| \frac{x - r}{w} \right\|_\infty ,$$

where

$$r_o = \frac{\sum_{i=1}^{n} \alpha_{oi} g_i}{\sum_{i=1}^{m} \beta_{oi} h_i} .$$

If $\gamma(\eta,s) = \frac{1+\eta}{2}$ (resp. $\gamma(\eta,s) = \frac{1-\eta}{2}$) ,

we have one-sided best rational Chebyshev approximation to x from

$$V^+ := \{r \in V \mid \forall_{s \in S} \ r(s) \geq x(s)\}$$

$$(\text{resp. } V^- := \{r \in V \mid \forall_{s \in S} \ r(s) \leq x(s)\}).$$

EXAMPLE 1.3. *Best approximation in a normed linear space.*

Let X be a normed linear space and let B_{X^*} be the unit ball in the continuous dual X^* of X. Let U_o be a non-empty open subset of $\mathbb{R}^{N-1}$ and let $B : U_o \longrightarrow X$ be a continuous mapping. For any $b \in X$, the set of best approximations to b from the set $V := B(U_1)$ is defined by

$$P_V(b) := \{v \in V \mid \|b - v\| = d(b,V)\} ,$$

where

$$d(b,V) := \inf \{\|b - v\| \in \mathbb{R} \mid v \in V\} .$$

An element

$$v_o := B(x_1^o, x_2^o, \ldots, x_{N-1}^o)$$

ist a best approximation of b from V if and only if

$$(x_1^o, x_2^o, \ldots, x_{N-1}^o, d(b,V))$$

is a minimum point of the following minimization problem:

MPA. Minimize $p(x_1, x_2, \ldots, x_N) := x_N$

subject to

$$\forall_{x^* \in B_{X^*}} \quad x^*(b(x_1, x_2, \ldots, x_{N-1})) - x_N \leq x^*(b).$$

Of course, the problem MPA is of type MP.

II. A CONDITION FOR OPTIMALITY

We denote by L the set of all parameters

$$\sigma \in C(T \times U) \times C(T) \times C(U),$$

for which the problem MP(σ) has a solution. We call this set the solvability set.

We state some definitions: For each parameter σ in L we define the set of feasible points

$$Z_\sigma := \bigcap_{t \in T} \{x \in U \mid A(t,x) \leq b(t)\},$$

the set of strict feasible points

$$Z_\sigma^< := \bigcap_{t \in T} \{x \in U \mid A(t,x) < b(t)\},$$

the minimum value

$$E_\sigma := \inf \{p(v) \in \mathbb{R} \mid v \in Z_\sigma\},$$

and the set of minimal solutions

$$P_\sigma := \{v \in Z_\sigma \mid p(v) = E_\sigma\}.$$

For each element $v_o \in U$ define the set

$$M_{\sigma,v_o} := \{t \in T \mid A(t,v_o) \geq b(t)\}.$$

Further let t_o be any point not in T and let $T_o := T \cup \{t_o\}$ be the compact Hausdorff-space with t_o as an isolated point. Then we extend the mapping $A : T \times U \longrightarrow \mathbb{R}$ to a continuous mapping $A : T_o \times U \longrightarrow \mathbb{R}$ by setting

$$\forall_{u \in U} \; A(t_o,u) := p(u).$$

KRABS [4,p.163] proved the following optimality condition:

<u>THEOREM 2.1.</u> Assume $\overline{Z_\sigma^<} = Z_\sigma$. Let v_o be an element in Z_σ such that

$$\forall_{u \in U} \; \min_{t \in M_{\sigma,v_o} \cup \{t_o\}} A(t,v_o) - A(t,u) \leq 0.$$

Then v_o is a minimal point.

For completeness we give the short proof. In fact, if v_o is not a minimal point, then there is an element v in $Z_\sigma^<$ such that

$$p(v_o) > p(v)$$

and

$$\forall_{t \in M_{\sigma,v_o}} \quad A(t,v_o) = b(t) > A(t,v) \ ,$$

which contradicts the assumption. □

III. THE CONDITION $\overline{Z_\sigma^<} = Z_\sigma$.

If the condition $\overline{Z_\sigma^<} = Z_\sigma$ is fulfilled, then the parameter σ satisfies the Slater condition, i.e. $Z_\sigma^< \neq \emptyset$. The converse is not true, in general. In the case of the minimization problem of example 1.3 the Slater condition is always fulfilled. We prove that we also have $\overline{Z_\sigma^<} = Z_\sigma$:

THEOREM 3.1. Let the minimization problem of example 1.3 be given. Then we have

$$\overline{Z_\sigma^<} = Z_\sigma .$$

PROOF. Let (v_o,z_o) be an element in $Z_\sigma \backslash Z_\sigma^<$. For each $\lambda > 0$ the element $(v_o,z_o + \lambda)$ is contained in $Z_\sigma^<$ and satisfies the inequality

$$\| (v_o,z_o) - (v_o,z_o + \lambda) \|_2 \leq \lambda ,$$

i.e. $\overline{Z_\sigma^<} = Z_\sigma$. □

For the further investigation we introduce the notion of a regular mapping.

DEFINITION 3.2. Let $\tilde{T}$ be a non-empty closed subset of T_o. A mapping $A : \tilde{T} \times U \longrightarrow \mathbb{R}$ is called regular if for each $\lambda > 0$, for each pair of elements v, v_o in U and for each closed subset $M \subset \tilde{T}$ such that

$$\forall_{t \in M} \quad A(t,v_o) - A(t,v) > 0$$

there exists an element $v_\lambda \in U$ such that $\|v_o - v_\lambda\|_2 < \lambda$ and

$$\forall_{t \in M} \; A(t,v_o) - A(t,v_\lambda) > 0.$$

REMARK. Each regular mapping is also pointwise convex (compare BROSOWSKI [1,p.58] . In general, the converse is not true. Now we can prove the

THEOREM 3.3. Let $\sigma = (A,b,p)$ be a parameter such that the mapping $A : T \times U \longrightarrow \mathbb{R}$ is regular and $Z_\sigma^< \neq \emptyset$. Then we have $\overline{Z_\sigma^<} = Z_\sigma$.

PROOF. Let v_o be an element in $Z_\sigma \setminus Z_\sigma^<$. We have show that for each $\lambda > 0$ there exists an element $v_\lambda \in Z_\sigma^<$ such that $\|v_o - v_\lambda\|_2 < \lambda$. It suffices to prove this for all sufficiently small $\lambda > 0$.

Since $v_o \in Z_\sigma \setminus Z_\sigma^<$, the set M_{σ,v_o} is non-empty and

$$\forall_{t \in M_{\sigma,v_o}} \; A(t,v_o) - A(t,v) > 0,$$

where v denotes an element in $Z_\sigma^<$. By compactness of M_{σ,v_o} and by continuity of A, there is a real number $\alpha > 0$ such that

$$\forall_{t \in M_{\sigma,v_o}} \; A(t,v_o) - A(t,v) \geq \alpha > 0.$$

Now define the open set

$$W := \{t \in T \mid A(t,v_o) - A(t,v) > \frac{\alpha}{2}\}.$$

By compactness of $T \setminus W$, one has

$$\forall_{t \in T \setminus W} \; A(t,v_o) - b(t) \leq -K < 0$$

with a suitable number $K > 0$. According to the definition of regularity, we can choose for any $\lambda > 0$ an element v_λ in U such that

$$A(t,v_o) - A(t,v_\lambda) > 0.$$

By the continuity of A, we can determine v_λ in U such that $\|v_o - v_\lambda\|_2 < \lambda$ and

$$\forall_{t \in T} \; A(t,v_\lambda) - A(t,v_o) \leq \lambda.$$

We can assume that $0 < \lambda < \frac{K}{2}$. If t in W, then we have the estimate

$$A(t,v_\lambda) = A(t,v_o) - (A(t,v_o) - A(t,v_\lambda))$$
$$< A(t,v_o) \leq b(t).$$

If $t \notin W$, then we have the estimate

$$A(t,v_\lambda) = A(t,v_o) - (A(t,v_o) - A(t,v_\lambda))$$
$$\leqq b(t) - K + \frac{K}{2} < b(t).$$

Consequently, v_λ is in $Z_\sigma^<$ and satisfies the inequality $\|v_o - v_\lambda\|_2 < \lambda$. □

IV. EXAMPLES OF REGULAR MAPPINGS

EXAMPLE 4.1. We use the notation of Example 1.1 and prove that the mapping

$$A : T_o \times \mathbb{R}^N \longrightarrow \mathbb{R}$$

is regular. To prove this, let be given: $\lambda > 0$, $v, v_o \in \mathbb{R}^N$ such that $A(t,v_o) - A(t,v) > 0$ for each t in a given closed subset $M \subset T_o$. Define the element

$$v_\lambda := (1 - \lambda\ )v_o + \lambda_1 v$$

with

$$0 < \lambda < \min(1, \frac{\lambda}{\|v - v_o\|_2}).$$

Then,

$$A(t,v_o) - A(t,v_\lambda) = \lambda_1 \langle B(t), v_o - v\rangle$$
$$= \lambda_1 (A(t,v_o) - A(t,v)) > 0$$

for each $t \in M$, and

$$\|v_o - v_\lambda\|_2 = \lambda_1 \|v_o - v\|_2 < \lambda.$$

Example 4.2. The mapping $A : T_o \times U \longrightarrow \mathbb{R}$ defined by

$$A(t,v) := \frac{\langle B(t),v\rangle}{\langle C(t),v\rangle}$$

is regular (for the notation compare example 1.2).

In fact, let $\lambda > 0$, $v_o, v \in U$ and a closed subset $M \subset T_o$ be given such that

$$\forall_{t \in M} \; A(t,v_o) - A(t,v) > 0.$$

Choose λ_1 such that

$$0 < \lambda_1 < \min(1, \frac{\lambda}{\|v - v_o\|_2})$$

and define the element

$$v_\lambda := \lambda_1 v + (1 - \lambda_1) v_o.$$

It follows that $\|v_o - v_\lambda\|_2 = \lambda_1 \|v_o - v_\lambda\|_2 < \lambda$.

For each $t \in M$ we have

$$A(t,v_o) - A(t,v_\lambda) = \lambda_1 \frac{<C(t),v>}{<C(t),v_\lambda>} \times$$

$$\times \left[\frac{<B(t),v_o>}{<C(t),v_o>} - \frac{<B(t),v>}{<C(t),v>} \right] > 0.$$

Consequently, the mapping $A : T_o \times U \longrightarrow \mathbb{R}$ is regular.

<u>EXAMPLE 4.3.</u> We use the notation of Example 1.2 and define a mapping

$$A : T_o \times U \times \mathbb{R} \longrightarrow \mathbb{R}$$

by setting

$$A(t,v,z) := \begin{cases} z & \text{if } t = t_o \\ \frac{<B(t),v>}{<C(t),v>} - \gamma(t)z & \text{if } t \in T. \end{cases}$$

Using the same method like in [3] we show that A is regular. In fact, let $\lambda > 0$, (v_o,z_o), $(v,z) \in U_o \times \mathbb{R}$, and a closed subset $M \subset T_o$ be given such that

$$\forall_{t \in M} \; A(t,v_o,z_o) - A(t,v,z) > 0.$$

The last inequality implies

$$\frac{<B(t),v_o>}{<C(t),v_o>} - \frac{<B(t),v>}{<C(t),v>} > \gamma(t)(z_o - z) \geq 0$$

for each $t \in M$. Choose λ_1 such that

$$0 < \lambda_1 < \min(1, \frac{\lambda}{2\|v - v_o\|_2})$$

and define the element

$$v_\lambda := \lambda_1 v + (1 - \lambda_1) v_o ,$$

which is contained in U_o.

CASE 1. $z_o - z > 0$, choose an ε such that

$$0 < \varepsilon < \min(\lambda_1 K, \frac{\lambda}{2(z_o - z)}) ,$$

where

$$K := \min_{t \in T} \frac{\langle C(t), v\rangle}{\langle C(t), v_\lambda\rangle} > 0,$$

and define $z_\lambda := z_o - \varepsilon(z_o - z)$.

It follows that

$$\begin{aligned} \|(v_o, z_o) - (v_\lambda, z_\lambda)\|_2 &\le \|v_o - v_\lambda\|_2 + |z_o - z_\lambda| \\ &= \lambda_1 \|v_o - v\|_2 + \varepsilon |z_o - z| \\ &< \frac{\lambda}{2} + \frac{\lambda}{2} = \lambda. \end{aligned}$$

For each $t \in M \setminus \{t_o\}$ we have:

$$\begin{aligned} &A(t, v_o, z_o) - A(t, v_\lambda, z_\lambda) \\ &= \lambda_1 \cdot \frac{\langle C(t), v\rangle}{\langle C(t), v_\lambda\rangle} \left[\frac{\langle B(t), v_o\rangle}{\langle C(t), v_o\rangle} - \frac{\langle B(t), v\rangle}{\langle C(t), v\rangle} \right] - \gamma(t)(z_o - z_\lambda) \\ &> \lambda_1 \cdot \frac{\langle C(t), v\rangle}{\langle C(t), v_\lambda\rangle} \gamma(t)(z_o - z) - \gamma(t)\varepsilon(z_o - z) \\ &\ge \gamma(t)(z_o - z)(\lambda_1 K - \varepsilon) > 0. \end{aligned}$$

If $t_o \in M$, then we have

$$A(t, v_o, z_o) - A(t, v_\lambda, z_\lambda) = \varepsilon(z_o - z) > 0.$$

CASE 2. $z_o - z \le 0$. In this case we have $t_o \notin M$. Define $z_\lambda := z_o$. It follows that

$$\|(v_o, z_o) - (v_\lambda, z_\lambda)\|_2 \le \|v_o - v_\lambda\|_2 < \frac{\lambda}{2} < \lambda.$$

For each $t \in M$ we have

$$A(t, v_o, z_o) - A(t, v_\lambda, z_\lambda)$$

$$= \lambda_1 \cdot \frac{<C(t)v>}{<C(t),v_\lambda>} \times \left[\frac{<B(t),v_o>}{<C(t),v_o>} - \frac{<B(t),v>}{<C(t),v>}\right] > 0.$$

Hence A is a regular mapping. □

V. THE NECESSITY OF THE CRITERION

In general the sufficient criterion of theorem 2.1 is not necessary for a minimal point. But, if the mapping $A : T_o \times U \longrightarrow \mathbb{R}$ is regular, then the criterion is also necessary. We first prove the

LEMMA 5.1. Let the minimization problem MP(σ) be given and suppose that the mapping

$$A : T_o \times U \longrightarrow \mathbb{R}$$

is regular. Let the elements v_o in Z_σ and $v \in U$ satisfy the inequality

$$\forall_{t \in M_{\sigma,v_o} \cup \{t_o\}} \quad A(t,v_o) - A(t,v) > 0.$$

Then for each $\lambda > 0$, there exists an element v_λ in Z_σ such that $p(v_\lambda) < p(v_o)$ and $\|v_o - v_\lambda\|_2 < \lambda$.

PROOF. *Case 1.* $M_{\sigma,v_o} = \emptyset$. Then there exists a constant $K > 0$ such that

$$\forall_{t \in T} \quad A(t,v_o) - b(t) \leq - K < 0.$$

Further we have by assumption $p(v_o) > p(v)$. By regularity of the mapping $A : T_o \times U \longrightarrow \mathbb{R}$ there exists a $v_\lambda \in U$ such that

$$\|v_o - v_\lambda\|_2 < \lambda \text{ and } p(v_o) - p(v_\lambda) > 0.$$

One can choose $\lambda > 0$ so small that

$$\|A(.,v_\lambda) - A(.,v_o)\|_\infty < \frac{K}{2}..$$

Then we have

$$\begin{aligned} A(t,v_\lambda) &= A(t,v_o) + (A(t,v_\lambda) - A(t,v_o)) \\ &\leq A(t,v_o) + \frac{K}{2} \\ &\leq b(t) - K + \frac{K}{2} < b(t) , \end{aligned}$$

i.e. $v_\lambda \in Z_\sigma$.

Case 2. $M_{\sigma,v_o} \neq \emptyset$. It suffices to prove the lemma for all sufficiently small $\lambda > 0$. By compactness of M_{σ,v_o} and by continuity of A, there is a real number $\alpha > 0$ such that

$$\underset{t\in M_{\sigma,v_o}}{\forall} \quad A(t,v_o) - A(t,v) \geq \alpha > 0$$

and $p(v_o) - p(v) \geqq \alpha$. Define the open set

$$W := \{t \in T \mid A(t,v_o) - A(t,v) > \frac{\alpha}{2}\}.$$

By compactness of T\W one has

$$\underset{t\in T\setminus W}{\forall} \quad A(t,v_o) - b(t) \leq - K < 0$$

with a suitable number $K > 0$. According to the definition of regularity, we can choose for any $\lambda > 0$ an element $v_\lambda \in U$ such that $p(v_o) - p(v_\lambda) > 0$ and

$$\underset{t\in \overline{W}}{\forall} \quad A(t,v_o) - A(t,v_\lambda) > 0.$$

By the continuity of A, we can determine v_λ in U such that $\|v_o - v_\lambda\|_2 < \lambda$ and

$$\underset{t\in T}{\forall} \quad A(t,v_\lambda) - A(t,v_o) \leq \lambda.$$

We can assume $0 < \lambda < \frac{K}{2}$. If t in W, then we have the estimate

$$A(t,v_\lambda) = A(t,v_o) - (A,(t,v_o) - A(t,v_\lambda))$$

$$\leq A(t,v_o) \leq b(t).$$

If $t \notin W$, then we have the estimate

$$A(t,v_\lambda) = A(t,v_o) - (A(t,v_o) - A(t,v_\lambda))$$

$$\geq b(t) - K + \frac{K}{2} < b(t).$$

Consequently, v_λ is in Z_σ and satisfies the relation $\|v_o - v_\lambda\|_2 < \lambda$. □

<u>THEOREM 5.2</u> Let $A : T_o \times U \longrightarrow \mathbb{R}$ be a regular mapping. Then for each $\sigma = (A\ ,b,p)$ in $\mathcal{L}$, we have: If v_o is in P_σ, then

$$\underset{u\in U}{\forall} \ \underset{t\in M_{\sigma,v_o}\cup\{t_o\}}{\min} \quad A(t,v_o) - A(t,u) \leq 0.$$

<u>PROOF.</u> Assume that there exists an element v in U such that

$$\forall_{t \in M_{\sigma,v_o} \cup \{t_o\}} \quad A(t,v_o) > 0.$$

By lemma 5.1, there exists an element v_λ in Z_σ such that $p(v_\lambda) < p(v_o)$ and consequently $v_o \notin P_\sigma$. □

VI. STRUCTURE OF THE SOLVABILITY SET $\mathsf{L}_{A,p}$

In the following we consider the case of a variable restriction function. In this case the solvability set can be written as

$$\mathsf{L}_{A,p} := \{b \in C(T) \mid (A,b,p) \in \mathsf{L}\}.$$

For each $\lambda \geq 0$ and for each $v_o \in U$ we define the continuous function

$$b_\lambda := b_{\lambda,v_o} := A(.,v_o) + \lambda(b - A(.,v_o)).$$

Now we prove

THEOREM 6.1. Let v_o be in P_b. Then we have:

(1) If $0 \leq \lambda_1 < \lambda_2$, then $v_o \in Z_{b_{\lambda_1}} \subset Z_{b_{\lambda_2}}$

and $Z^<_{b_{\lambda_1}} \neq \emptyset \Rightarrow Z^<_{b_{\lambda_2}} \neq \emptyset$;

(2) $\forall_{0 \leq \lambda \leq 1} \; b_\lambda \in \mathsf{L}_{A,p}$;

(3) If $A : T_o \times U \longrightarrow \mathbb{R}$ is regular and $Z^<_b \neq \emptyset$,

then

$\forall_{\lambda > 1} \; b_\lambda \in \mathsf{L}_{A,p}$;

(4) If $\forall_{\lambda > 1} \; v_o \in P_{b_\lambda}$, then

$$(*) \quad \forall_{u \in U} \; \min_{t \in M_{b,v_o} \cup \{t_o\}} A(t,v_o) - A(t,u) \leq 0.$$

PROOF. (1) Easy to see.

(2) Immediate consequence of (1).

(3) Since A is regular, we have

$$\underset{u\in U}{\forall}\ \min_{t\in M_{b,v_o}\cup\{t_o\}} A(t,v_o) - A(t,u) \leq 0.$$

Property (1) implies $Z^{<}_{b_\lambda} \neq \emptyset$ for each $\lambda > 1$. By Theorem 3.3 we have

$$\overline{Z^{<}_{b_\lambda}} = Z_{b_\lambda} .$$

Since $M_{b_\lambda, v_o} = M_{b,v_o}$, it follows also

$$\underset{u\in U}{\forall}\ \min_{t\in M_{b_\lambda,v_o}\cup\{t_o\}} A(t,v_o) - A(t,u) \leq 0.$$

Then Theorem 2.1 implies $v_o \in P_{b_\lambda}$, i.e. $b_\lambda \in L_{A,p}$.

(4) The case $M_{b,v_o} = M_{b_\lambda,v_o} = \emptyset$ is trivial. Assume that $M_{b,v_o} = M_{b_\lambda,v_o}$ $\neq \emptyset$ and that (*) is not true. Then there exist an element $u \in U$ and a real number α such that

$$\underset{t\in M_{b,v_o}\cup\{t_o\}}{\forall} A(t,v_o) - A(t,u) \geq \alpha > 0.$$

The open set

$$W := \{t \in T \mid A(t,v_o) - A(t,u) > \tfrac{\alpha}{2}\}$$

contains M_{b,v_o}. Consequently there is a real number $K > 0$ such that

$$\underset{t\in T\setminus W}{\forall} A(t,v_o) - b(t) \leq -K < 0.$$

Choose a real number λ such that

$$\lambda > \max(1, \frac{\|A(.,v_o) - A(.,u)\|_\infty}{K}).$$

If $t \in W$, then we have

$$\begin{aligned} A(t,u) - b_\lambda(t) &= A(t,u) - A(t,v_o) - \lambda(b(t) - A(t,v_o)) \\ &< -\frac{\alpha}{2} < 0. \end{aligned}$$

If $t \in T\setminus W$, then we have

$$\begin{aligned} A(t,u) - b_\lambda(t) &\leq \|A(.,u) - A(.,v_o)\|_\infty - \lambda(b(t) - A(t,v_o)) \\ &< \lambda K - \lambda K = 0. \end{aligned}$$

Consequently $u \in Z_{b_\lambda}$. Since $p(v_o) > p(u)$ it follows that $v_o \notin P_{b_\lambda}$, a contradition. □

VII. CHARACTERIZATION OF A SET OF MINIMAL POINTS

In this section we use Theorem 6.1 to derive a characterization of a set of minimal points. The proof is with a minor modification the same like in [3].

Let V be a subset of Z_b. Then we define

$$M_{b,V} := \bigcap_{v\in V} M_{b,v}.$$

THEOREM 7.1. Let $A : T_o \times U \longrightarrow \mathbb{R}$ be a regular mapping and let $b \in \mathsf{L}_{A,p}$ such that $Z_b^< \neq \emptyset$. Then a subset $V \subset Z_b$ is a set of minimal points for MP(b) if and only if.

$$\forall_{v_o\in V} \ \forall_{v\in U} \ \min_{t\in M_{b,V}\cup\{t_o\}} A(t,v_o) - A(t,v) \leq 0.$$

PROOF. Sufficiency follows from Theorem 2.1 and Theorem 3.3., since $M_{b,V} \subset M_{b,v_o}$ for each $v_o \in V$.

To prove the necessity, we consider first the case where V contains only a finite number n of elements and proceed by induction. If $n = 1$, then $M_{b,V} = M_{b,v_o}$ and Theorem 5.2 applies. Let $v_1, v_2, \dots, v_{n+1}$ be minimal points for MP(b) and v in U be given. If $p(v_\nu) - p(v) \leq 0$ for some ν with $1 \leq \nu \leq n+1$, then $p(v_\nu) - p(v) \leq 0$ for all $1 \leq \nu \leq n+1$. Thus, we need only consider the case $p(v_\nu) - p(v) > 0$ for all $1 \leq \nu \leq n+1$. For $\lambda > 1$ consider the function

$$b_\lambda := A(.,v_{n+1}) + \lambda(b - A(.,v_{n+1})).$$

Theorem 6.1. implies $v_{n+1} \in P_{b_\lambda}$. For $\nu = 1,2,\dots,n$ and $t \in T$, we have the inequality

$$\begin{aligned} A(t,v_\nu) - b_\lambda(t) &= A(t,v_\nu) - A(t,v_{n+1}) - \lambda(b(t) - A(t,v_{n+1})) \\ &= A(t,v_\nu) - b(t) - (\lambda-1)(b(t) - A(t,v_{n+1})) \\ &\leq A(t,v_\nu) - b(t) \leq 0, \end{aligned}$$

which implies $v_1, v_2, \dots, v_n \in Z_{b_\lambda}$ and consequently $v_1, v_2, \dots, v_n \in P_{b_\lambda}$. By induction hypothesis the set

$$M_n := \bigcap_{\nu=1}^{n} M_{b_\lambda, v_\nu}$$

satisfies the inequality

$$\underset{v \in U}{\forall} \quad \min_{t \in M_n \cup \{t_o\}} \quad A(t,v_\nu) - A(t,v) \leqq 0.$$

Then there exists a point t in M_n such that

$$A(t,v_\nu) - b_\lambda(t) = 0 \quad \& \quad A(t,v_\nu) - A(t,v) \leqq 0 \, ,$$

$\nu = 1,2,\ldots,n$. Thus we have

$$A(t,v_\nu) - A(t,v_{n+1}) - \lambda(b(t) - A(t,v_{n+1})) = 0$$

or

$$[A(t,v_\nu) - b(t)] \; + (\lambda - 1) \; [A(t,v_{n+1}) - b(t)] \quad = 0,$$

which implies $A(t,v_\nu) = b(t)$ for $\nu = 1,2,\ldots,n+1$, and

$$t \in M_{n+1} := \bigcap_{\nu=1}^{n+1} M_{b,v_\nu}.$$

Moreover, $A(t,v_\nu) - A(t,v) \leqq 0$ for $\nu = 1,2,\ldots,n$ and $t \in M_{n+1}$ imply $A(t,v_{n+1}) - A(t,v) \leqq 0$, which concludes the proof for V finite.

To prove the general case, we assume that there exist elements $v_o \in V$ and $u \in U$ such that

$$\underset{t \in M_{\sigma,V} \cup \{t_o\}}{\forall} \qquad A(t,v_o) - A(t,u) > 0.$$

This inequality implies

$$\bigcap_{v \in V} (M_{b,v} \cup \{t_o\}) \;\; \cap \; \{t \in T_o \,|\, A(t,v_o) - A(t,u) \leqq 0\} = \emptyset.$$

By compactness of the sets

$$(M_{b,v} \cup \{t_o\}) \quad \cap \; \{t \in T_o \,|\, A(t,v_o) - A(t,u) \leqq 0\} \, ,$$

there exist elements $v_1,v_2,\ldots,v_n$ in V such that

$$\bigcap_{\nu=1}^{n} (M_{b,v_\nu} \cup \{t_o\}) \; \cap \quad \{t \in T_o | A(t,v_o) - A(T,u) \leqq 0\} = \emptyset \, ,$$

which is impossible by the first part. □

VIII. THE UPPER SEMICONTINUITY OF THE MINIMUM SET

THEOREM 8.1. Let $L_{A,p}$ be open and let $P : L_{A,p} \longrightarrow POT(U)$ be upper semicontinuous and compact valued, and let $Q_b := A(.,P_b)$ be convex for each b in $L_{A,p}$.

Then, for each b in $L_{A,p}$, there exists an element v_o on P_b, which satisfies the criterion

$$\forall_{u\in U} \min_{t\in M_{b,v_o}\cup\{t_o\}} A(t,v_o) - A(t,u) \leq 0.$$

REMARK. The proof is similar to the proof of theorem 4.1 in [2] .

PROOF. By theorem 6.1,(4) it suffices to show, that for each b in $L_{A,p}$ there is an element v_o in P_b, which satisfies the condition

$$\forall_{\lambda>1} \quad v_o \in P_{b_\lambda} ,$$

where $b_\lambda := A(.,v_o) + \lambda(b - A(.,v_o))$.

If this were false, one could choose an element b in $L_{A,p}$, such that

$$\forall_{v\in P_b} \quad \exists_{\lambda>0} \quad v \notin P_{b_\lambda} .$$

By the upper semicontinuity of P, for each $v \in P_b$, there exists a $\lambda(v) \geq 0$ such that

(*) $$\forall_{\lambda\in[0,\lambda(v)]} [v \in P_{b_\lambda} \quad \& \quad \forall_{\lambda>\lambda(v)} \quad v \notin P_{b_\lambda}].$$

CASE 1. *The set* $\{\lambda(v) \in \mathbb{R} \mid v \in P_b\}$ *is unbounded.*

Then, for each $n \in \mathbb{N}$, there exists an element v_n in P_b such that

$$\forall_{n\in\mathbb{N}} \quad v_n \in P_{b_n} ,$$

where $b_n := A(.,v_n) + n(b - A(.,v_n))$. By compactness of P_b, the sequence (v_n) has an accumulation point v_o in P_b. We claim, that for the element v_o there does not exist a $\lambda(v_o)$ with the property (*).

For the proof, let $\lambda > 0$ be arbitrary. Then the element

$$b_\lambda := A(.,v_o) + \lambda(b - A(.,v_o))$$

is an accumulation point of the sequence

$$g_n := A(.,v_n) + \lambda(b - A(.,v_n)).$$

By theorem 6.1,(2) the element v_n is contained in P_{g_n} for almost all $n \in \mathbb{N}$. By the upper semicontinuity of P, it follows that v_o in P_{b_λ}. Since $\lambda > 0$ was chosen arbitrarily, there does not exist a $\lambda(v_o)$ with property (*), consequently, we have

CASE 2. *The set* $\{\lambda(v) \in \mathbb{R} \mid v \in P_b\}$ *is bounded.*

Define $\lambda_o := \sup \{\lambda(v) \in \mathbb{R} \mid v \in P_b\}$. By the upper semicontinuity of P, there exists an element v_o in P_b such that $v_o \in P_g$, where

$$g := A(.,v_o) + \lambda_o(b - A(.,v_o)).$$

By compactness of P_g, there exists a compact neighborhood $W \subset U$ of P_g. Then the set

$$\overline{\text{con}(A(.,W))}$$

is also compact. By the upper semicontinuity of P there exists a bounded and convex neighborhood V of g such that

$$\forall_{h \in V} \quad P_h \subset W.$$

By compactness of $\overline{\text{con}(A(.,W))}$, we can find a real number $r > 0$ such that the compact and convex set

$$K_r := g + \frac{r}{r+1} [g - \overline{\text{con}(A(.,W))}]$$

is contained in V. The set-valued mapping

$$\Lambda(h) := g + \frac{r}{r+1} [g - A(.,P_h)]$$

is upper semicontinuous and compact- and convex-valued. Since $K_r \subset V$, the compact and convex set K_r is mapped by Λ into $POT(K_r)$. By the fixed point theorem of KY FAN, this mapping has a fixed point, i.e. there exists an element h in K_r with $h \in \Lambda(h)$, which implies that there is an element v_1 in P_h such that

$$h = g + \frac{r}{r+1} [g - A(.,v_1)]$$

or

$$g = \frac{1+r}{1+2r} h + \frac{r}{1+2r} A(.,v_1).$$

Consequently, g is contained in the interior of the segment $[A(.,v_1),h]$. By theorem 6.1.,(2) the element v_1 is contained in P_g. Since we have $Z_h \supset Z_g$ and $v_o \in Z_g$, it follows that $v_o \in P_h$. The elements v_o, v_1, g, h satisfy the relation

$$A(.,v_o) = g + \frac{\lambda_o}{\lambda_o - 1}(b - g).$$

Now we can determine real numbers τ and ρ and an element $\overline{v}$ in P_h such that

$$h = A(.,\overline{v}) + \tau(b - A(.,\overline{v}))$$

and

$$A(.,\overline{v}) = \rho A(.,v_1) + (1-\rho)A(.,v_o).$$

The computation yields

$$\tau = \lambda_o(1 + r) > \lambda_o \qquad \rho = \frac{r}{\lambda_o(r+1)-1} < 1.$$

By convexity of $A(.,P_h)$, the element $\overline{v}$ is contained in P_h. Since $\lambda_o(1+r) > \lambda_o$, by theorem 6.1,(2), it follows that $\overline{v} \in P_b$. Consequently, λ_o is not the supremum of the set

$$\{\lambda(v) \in \mathbb{R} \mid v \in P_b\},$$

which is a contradition. □

[1] B.BROSOWSKI, Parametric semi- infinite optimization, Lang,Frankfurt(M) and Bern, 1982.

[2] B.BROSOWSKI, Parametric approximation and optimization, in: Functional analysis, holomorphy, and approximation theory,pp. 93 - 116, North-Holland, Amsterdam, 1982.

[3] B.BROSOWSKI and C.GUERREIRO, On the characterization of a set of optimal points and some applications, in:Approximation and Optimization in Mathematical Physics, pp.141 - 174, Lang, Frankfurt(M), Bern, 1983.

[4] W.KRABS, Optimierung und Approximation, B.G.Teubner, Stuttgart, 1975.

STRUCTURE THEORY FOR REALIZATIONS OF FINITE VOLTERRA SERIES

P.E. Crouch and P.C. Collingwood.
Control Theory Centre,
University of Warwick,
Coventry, CV4 7AL,
England.

1. INTRODUCTION

An input-output map described by a finite Volterra series of length p has the form

$$y_j(t) = W_{jo} + \int_o^t \sum_{i_1} W_{j1}^{i_1}(t,\sigma_1)u_{i_1}(\sigma_1)d\sigma_1 + \ldots \qquad -1$$

$$+ \int_o^t \int_o^{\sigma_{p-1}} \sum_{i_1\ldots i_p} W_{jp}^{i_1\ldots i_p}(t,\sigma_1\ldots\sigma_p)u_{i_1}(\sigma_1)\ldots u_{i_p}(\sigma_p)d\sigma_1\ldots d\sigma_p.$$

$$1 \leq i_j \leq m, \quad 1 \leq j \leq q$$

We shall only be concerned with those input-output maps defined by Volterra series in which the Volterra kernels, abbreviated to $W_k(t\sigma_1\ldots\sigma_k)$, (i) are continuously differentiable, satisfy (ii) $W_o(t) \equiv 0$ and satisfy

$$\text{(iii)}\left(\frac{\partial}{\partial t} + \sum_{i=1}^{k}\frac{\partial}{\partial\sigma_i}\right) W_k(t,\sigma_1\ldots\sigma_k) \equiv 0 \qquad 1 \leq k \leq p.$$

The input-output map is then stationary (is invariant under time translation). We shall refer to the input-output map defined by such a Volterra series, by the words **stationary** finite Volterra series abbreviated s.f.v.s.

The first fundamental result concerns the realizability of s.f.v.s. by finite dimensional nonlinear systems.

THEOREM 1 BROCKETT [1]

A s.f.v.s. has a realization by a nonlinear system

$$\dot{x} = f(x) + \sum_{i=1}^{m} u_i g_i(x), \quad x(o) = x_o, \quad f(x_o) = 0 \qquad -2$$

$$y_i = h_i(x), \quad 1 \leq i \leq q$$

where $x \in M$ is a real analytic manifold, f, $g_1\ldots g_m$ are analytic vector fields on M, h_i, $1 \leq i \leq q$ are analytic functions on M, if and only if it has a realization by a bilinear system

$$\dot{x} = Ax + \sum_{i=1}^{m} u_i N_i x, \quad x \in \mathbb{R}^n, \quad Ax_o = 0, \quad x(o) = x_o$$

-3

$$y_i = c_i' x$$

In this case the Volterra kernels W_k are analytic on $t \geq \sigma_1 \ldots \geq \sigma_k \geq 0$, and hence define analytic functions on the whole of $\mathbb{R}^{k+1}$. Such realizations exist if and only if each Volterra kernel is differentiably separable. That is for each kernel W_k there exists an integer N and continuously differentiable functions γ_i^j such that $W_k(t,\sigma_1 \ldots \sigma_k) = \sum_{j=1}^{N} \gamma_o^j(t)\gamma_1^j(\sigma_1)\ldots\gamma_k^j(\sigma_k)$.

Bilinear realizations as in equation (3) will never be minimal in the sense of Sussmann [10] (that is orbit minimal and observable). This presents the problem of constructing a class of canonical realization in the form of systems (2) which have input-output maps described by s.f.v.s., and in which minimal realizations of s.f.v.s. can always be found. Such a class of canonical realization should generalize the linear case where the linear system

$$\dot{x} = Ax + \sum_{i=1}^{m} b_i u_i, \quad x(o) = 0, \quad x \in \mathbb{R}^n.$$

$$y_i = c_i' x, \qquad 1 \leq i \leq q$$

is generally accepted as the canonical realization of a s.f.v.s. of length one.

Let L denote the Lie algebra of vector fields on M generated by f, $g_1 \ldots g_m$ in a system (2). Note that all vector fields in the Lie algebra of the bilinear system (3) are complete, and that this Lie algebra is finite dimensional. From the existence theorem for minimal realizations Sussmann [10], any minimal realization of a s.f.v.s. will have a Lie algebra which is also finite dimensional and consists of complete vector fields. By the uniqueness theorem for minimal realizations [10], this Lie algebra is unique up to isomorphism, and the state space in such a minimal realization is unique up to diffeomorphism.

Let H be the smallest linear space of functions on M containing $h_1 \ldots h_q$ and closed under Lie differentiation by elements of L, or simply f, $g_1 \ldots g_m$. We call H the observation space. The following result is a partial converse to theorem (1), and is a specialised version of a similar result for bilinear systems, Fliess and Kupka [5].

<u>THEOREM 2</u> <u>CROUCH AND COLLINGWOOD</u> [3]

An analytic system (2) has a s.f.v.s. of length p if and only if

(i) H is finite dimensional

(ii) There exists a sequence of p distinct subspaces $\bar{H}^i \subset H$, $1 \leq i \leq p$,

$$\{0\} = \bar{H}^{p+1} \subset \bar{H}^{p} \subset \bar{H}^{p-1} \cdots \subset \bar{H}^{2} \subset \bar{H}^{1} \subset \bar{H}^{0} = H$$

such that $L_{g_i} \bar{H}^k \subset \bar{H}^{k+1}$, $\quad 0 \leq k \leq p+1$

$1 \leq i \leq m$

$L_f \bar{H}^k \subset \bar{H}^k$, $\quad 0 \leq k \leq p+1$

where $\bar{H}^p$ is the one dimensional space of constant functions, and there exist no sequence of less than p subspaces with the same properties.

(L_X refers to Lie Differentiation by the vector field X, and [X, Y] will denote the Lie bracket of two vector fields X and Y).

If a system (2) has a s.f.v.s. then there is a natural choice for the subspace $\bar{H}^k$ in the above result, namely the subspace spanned by the functions

$$L_f^{k_0} L_{g_{i_1}} L_f^{k_1} \cdots L_{g_{i_r}} L_f^{k_r} (h_j)$$

for $r \geq k$, $\quad 1 \leq i_\ell \leq m$, $\quad 1 \leq j \leq q$, $\quad k_\ell \geq 0$.

We shall therefore reserve the symbol H^k for this subspace of H.

The final result of this section gives further information on the state space and Lie algebra in minimal realization of s.f.v.s.

<u>THEOREM 3 CROUCH</u> [2]

Given a minimal realization of a s.f.v.s. of length p then (i) the state space is diffeomorphic to a vector space, (ii) the Lie algebra is solvable, and the ideal $S \subset L$, generated by $g_1 \ldots g_m$, is nilpotent with descending central series $\{S^i\}$, defined by $S^{k+1} = [S, S^k]$, satisfying

$$\{0\} = S^{p+1} \subset S^p \subset \ldots S^2 \subset S^1 = S$$

Any analytic realization (2) of a s.f.v.s. is minimal if and only if it is strongly accessible and locally weakly observable.

The latter part of this theorem shows that it is possible to check minimality of realizations of s.f.v.s. simply by checking the conditions

$$S(x) = T_x M, \quad dH(x) = T_x^* M, \quad \forall\, x \in M$$

where $S(x) = \{X(x);\ X \in S\}$, $dH(x) = \{dh(x),\ h \in H\}$ are distributions in the tangent and cotangent bundles TM and T^*M respectively (see Sussmann and Jurdjevic [11] and Hermann and Krener [8] for details concerning the concepts of strong accessibility and local weak observability).

The conditions above give the generalization of the rank conditions for linear systems required for this particular class of nonlinear system. They reduce to the rank conditions when p=1 so that we are dealing with linear systems. To check the conditions in general one must resort to coordinate charts to obtain a local coordinate representation of the system. However motivated by the first result in theorem (3) one expects to find global coordinate systems. The next section demonstrates that this can achieved yielding a rich algebraic strcuture for the systems obtained.

2. CANONICAL FORMS AND GRADED STRUCTURES

We introduce the notation and some results from the theory of graded vector spaces (see Goodman [6] for a good account of the theory). If

$n = \sum_{i=1}^{p} n_i$ then we may decompose $\mathbb{R}^n$ into a direct sum

If $x \in \mathbb{R}^n$ we then write it as $(x_1 \dots x_p)$ with $x_i \in \mathbb{R}^{n_i}$, and write $x_i = (x_i^1 \dots x_i^{n_i})$. We define for each $t > o$ a diffeomorphism δ_t of $\mathbb{R}^n$ by setting

$$\delta_t(x_1 \dots x_p) = (tx_1, t^2x_2, \dots t^p x_p)$$

We call the pair $(\mathbb{R}^n, \delta_t)$ a graded vector space of degree p.

Given a graded vector space of degree $p(\mathbb{R}^n, \delta_t)$ we make the following definitions. A polynomial function h on $\mathbb{R}^n$ is homogeneous of degree k if $h \circ \delta_t(x) = t^k h(x)$ for each $x \in \mathbb{R}^n$. Denote by H^k the vector space for all homogeneous polynomials of degree k on $\mathbb{R}^n$.

A vector field X on $\mathbb{R}^n$ with polynomial coefficients is said to be homogeneous of degree m, $m \leq p$, if for each $k \geq 0$ and each $h \in H^k$ we have $L_X(h) \in H^{k-m}$ (We set $H^k = \{0\}$ for $k < o$). In other words $\delta_{t*}X = t^m(X o \delta_t)$. We denote by P^m the vector space of all vector fields on $\mathbb{R}^n$ which are homogeneous of degree m.

We set $V^m = P^p \oplus \dots \oplus P^m$, $V^k = \{0\}$ for $k > p$ and $C^m = H^m \oplus \dots \oplus H^o$ with $C^k = \{0\}$ for $m < o$. A one form ω on $\mathbb{R}^n$ with polynomial coefficients is said to be homogeneous of degree k, if for each $m \geq o$ and vector field $X \in P^m$ we have $\omega(X) \in H^{k-m}$, or in other words $\delta_t^*\omega = t^k(\omega o\ \delta_t)$. The space of all closed and hence exact one forms on $\mathbb{R}^n$, homogeneous of degree k may be identified with dH^k. Set $dC^k = dH^k \oplus \dots \oplus dH^1$, since H^o consists of constant functions. We may easily show that $H^k \otimes P^m \subset P^{m-k}$, $H^k \otimes H^m \subset H^{k+m}$, $[P^k, P^m] \subset P^{m+k}$ and further if $h \in C^m$, $X \in V^k$ then $L_X(h) \in C^{m-k}_o$ and $[V^m, V^k] \subset V^{m+k}$.

We let W^k be the linear space of constant vector fields lying in P^k, so that W^k is spanned by $\partial/\partial x_k^i$ for $1 \leq i \leq n_k$. We let Z^k be the linear space of constant one forms lying in dH^k, so that Z^k is spanned by dx_k^i, $1 \leq i \leq n_k$. We may now easily establish the following facts.

$$P^j = \bigoplus_{k=j}^{p} (H^{k-j} \otimes W^k), \qquad V^j = \bigoplus_{k=j}^{p} (C^{k-j} \otimes W^k)$$

$$dH^j \subset \bigoplus_{k=1}^{j} (H^{j-k} \otimes Z^k), \qquad dC^j \subset \bigoplus_{k=1}^{j} (C^{j-k} \otimes Z^k) \qquad \text{for } 1 \leq j \leq p$$

and

$$P^o = \bigoplus_{k=1}^{p} (H^k \otimes W^k), \qquad V^o = \bigoplus_{k=1}^{p} (C^k \otimes W^k)$$

Consider now a system defined on a graded vector space $(\mathbb{R}^n, \delta_t)$ of degree p

$$\dot{x} = F(x) + \sum_{i=1}^{m} u_i G_i(x), \quad x(o) = 0, \quad x \in \mathbb{R}^n$$
$$F(0) = 0. \tag{4}$$
$$y_j = H_j(x)$$
$$F \in V^o, \quad G_i \in V^1, \quad 1 \leq i \leq m, \quad H_j \in C^p, \quad 1 \leq j \leq q.$$

Using the identities above we may write this system in terms of the expanded version of the state vector $x = (x_1, \ldots x_p)$ in the following way

$$\begin{aligned}
\dot{x}_1 &\quad A_1 x_1 & & b_{i_1} & x_1(o) = 0 \\
\dot{x}_2 &= A_2 x_2 + a_2(x_1) & + \sum_{i=1}^{m} u_i \quad & b_{i_2}(x_1) & x_2(o) = 0 \\
\vdots &\quad \vdots & & \vdots & \vdots \\
\dot{x}_p &\quad A_p x_p + a_p(x_1 \ldots x_{p-1}) & & b_{i_p}(x_1 \ldots x_{p-1}) & x_p(o) = 0
\end{aligned}$$

$$y_j = H_j(x_1 \ldots x_p) \tag{5}$$

where $a_j^k \in C^j$ and $b_{i_j}^k \in C^{j-1}$, $H_j \in C^p$. Here the polynomial components a_j^k contain no linear or constant terms.

Let H be the observation space for such a system then $H \subset C^p$ and is therefore a finite dimensional space. The spaces $\overline{H}^i = H \cap C^{p-i}$ for $0 \leq i \leq p+1$ satisfy the conditions (ii) of theorem (2), except that they may not be distinct or non zero. It follows that system (4) has an input-output map defined by a s.f.v.s. of length less than or equal to p. With this motivation we make the following definition.

A system (2) which is a realization of a s.f.v.s. of length p is in canonical form if it is defined on a graded vector space ($\mathbb{R}^n$, δ_t) of degree p and has the structure defined in (4).

Note that when p = 1 system (4) has the structure of the canonical linear system as desired.

We now show that a realization of a s.f.v.s. in canonical form has the Lie algebra structure described in theorem (3). Indeed in the situation described in (4) the Lie algebra $L \subset V^0$ and the ideal $S \subset V^1$. Let $\bar{S}^i = S \cap V^i \subset S$. Using the properties of V^i we see that $\bar{S}^{i+1}$ is an ideal in $\bar{S}^i$ such that $\bar{S}^i/\bar{S}^{i+1}$ is an abelian Lie algebra with $\bar{S}^{p+1} = \{0\}$. It follows that $\bar{S}$ is a nilpotent Lie algebra and since $S^i \subset S \cap V^i = \bar{S}^i$ we have $S^{p+1} = \{0\}$, so that the descending central series $\{S^i\}$ has length less than or equal to p.

Having described general canonical forms we specialize them in two ways.

<u>GRADED CONTROLLABLE CANONICAL FORM</u>

A system (2) which is a realization of a s.f.v.s. of length p is said to be in graded controllable canonical form (abbreviated to g.c.c.f.) if it is in canonical form on a graded vector space ($\mathbb{R}^n$, δ_t) of degree p, and in addition

$$V^i(o) = S^i(o), \quad 1 \leq i \leq p,$$

where $V^i(o)$ and $S^i(o)$ are the subspaces of $T_o \mathbb{R}^n$ spanned by the corresponding Lie algebras at O.

From the structure of V^i we see that we may write all elements in V^i in the form

$$\sum_k r_i^k \partial/\partial x_i^k + \sum_{j>i} r_j^k(x_1 \dots x_{j-i}) \partial/\partial x_j^k$$

Since for systems in canonical form $S^i \subset V^i$ it follows that a system in g.c.c.f. satisfies

$$V^i(x) = \bigoplus_{j=p}^{i} W^j(x) = S^i(x), \quad \forall x \in \mathbb{R}^n.$$
$$1 \leq i \leq p.$$

In particular systems in g.c.c.f. are strongly accessible.

The justification for making this definition lies in the following result

THEOREM 4 CROUCH [2]

A s.f.v.s. of length p which is realizable by an analytic system (2) is also realizable by a minimal analytic system in g.c.c.f. on a graded vector space $(\mathbb{R}^n, \delta^t)$ of degree p. The integers $n_i = \text{Dim.}(S^i(o)/S^{i+1}(o))$, $1 \leq i \leq p$, depend only on the Volterra series not on the choice of g.c.c.f. and define the graded structure. $\mathbb{R}^n = \mathbb{R}^{n_1} \oplus \mathbb{R}^{n_2} \oplus \dots\dots \oplus \mathbb{R}^{n_p}$.

We note that although we must have $n_1 \neq 0$ there is no such restriction on the integers n_i, $2 \leq i \leq p$. For example the system $\dot{x} = u$, $x(o) = 0$

$$y = x^p$$

has a Volterra series of length p and $n_1 = 1$, $n_i = o$, $2 \leq i \leq p$.

There is an algorithm for extracting the integers n_i directly from the Volterra kernels; see Crouch [2] for details.

GRADED OBSERVABLE CANONICAL FORM

A system (2) which is a realization of a s.f.v.s. of length p is said to be in graded observable canonical form (abbreviated to g.o.c.f.) if it is in canonical form on a graded vector space $(\mathbb{R}^n, \delta_t)$ of degree p, and in addition

$$d\,C^{p-i+1}(o) = d\,H^{i-1}(o), \quad 1 \leq i \leq p$$

where $d\,C^i(o)$ and $d\,H^i(o)$ are the subspaces of $T_o^*\mathbb{R}^n$ spanned by the corresponding spaces of one forms.

From the structure of $d\,C^{p-i+1}$ we may write all elements in the form

$$\sum_k r_i^k\, dx_{p-i+1}^k + \sum_{\substack{j < p-i+1 \\ k}} r_j^k\,(x_1 \dots x_{p-i+1-j})\, dx_j^k$$

Since for systems in canonical form $H^{i-1} \subset C^{p-i+1}$ it follows that a system in g.o.c.f. satisfies

$$d\,C^{p-i+1}(x) = \bigoplus_{j=1}^{p-i+1} Z^j(x) = d\,H^{i-1}(x) \quad \forall\, x \in \mathbb{R}^n, \quad 1 \leq i \leq p$$

In particular systems in g.o.c.f. are locally weakly observable. Other canonical forms, for arbitrary systems have been made using the structure of H, see for example Gauthier and Bornard [7] and Nijmeijer [9]. However these canonical forms have a different structure from the one given here, although their method of construction is similar to that in Crouch and Collingwood [3].

The justification for making the definition given here lies in the following result.

THEOREM 5 CROUCH AND COLLINGWOOD [3]

A s.f.v.s. of length p which is realizable by an analytic system (2) is also realizable by a minimal analytic system in g.o.c.f., on a graded vector space $(\mathbb{R}^n, \delta_t)$ of degree p. The integers $m_i = \text{Dim } dH^{p-i}(o)/dH^{p-i+1}(o)$ $1 \leq i \leq p$ depend only on the Volterra series not on the choice of g.o.c.f., and define the graded structure $\mathbb{R}^n = \mathbb{R}^{m_1} \oplus \mathbb{R}^{m_2} \oplus \ldots . \oplus \mathbb{R}^{m_p}$.

Here again although $m_1 \neq 0$ there is no restriction in the remaining integers m_i. One naturally asks whether one can obtain minimal realizations of s.f.v.s. which are simultaneously in g.c.c.f. and g.o.c.f., and hence the situation in which

$$m_i = n_i \qquad 1 \leq i \leq p.$$

In general this is not possible, as shown in the example below, taken from Crouch and Collingwood [3]. However there is a specific case where this situation does arise.

THEOREM 6 CROUCH AND COLLINGWOOD [3]

A s.f.v.s. defined by a single Volterra kernel W_p, realizable by an analytic system (2), is also realizable by a minimal analytic system which is simultaneously in graded controllable and graded observable canonical forms, with respect to the same graded vector space $(\mathbb{R}^n, \delta_t)$ of degree p.

Of course the special case of linear systems p = 1 is included in this result. As another non-trivial example of the situation in theorem (6) we may take the system

$$\dot{x}_1 = u \ , \qquad x_1(o) = 0$$
$$\dot{x}_2 = x_1^2 \ , \qquad x_2(o) = 0$$
$$y = x_2 \ .$$

This is minimal and has a Volterra series consisting of a single second order term. It is in g.o.c.f. and g.c.c.f. with respect to the graded structure $\mathbb{R}^2 = \mathbb{R} \oplus \mathbb{R}$ $\delta_t(x_1,x_2) = (tx_1,t^2x_2)$.

The main reason for the result in theorem (6) is that when one constructs a minimal realization of a s.f.v.s., defined by a single Volterra kernel W_p, in g.c.c.f. as in theorem (4), the resultant system may be written as in (4) but with $F \in P^o$, $G_i \in P^1$, $H_j \in H^p$, $1 \leq i \leq m$, $1 \leq j \leq q$. As a result in the representation given in (5) we now have $a_j^k \in H^j$ and $b_{ij}^k \in H^{j-1}$. For details see Crouch [2].

Finally in this section we give an example of a situation in which a simultaneous graded observable and graded controllable canonical structure cannot be found.

The system
$$\begin{aligned} \dot{x}_1 &= u, & x_1(o) &= 0, \\ \dot{x}_2 &= x_1^2, & x_2(o) &= 0, \\ \dot{x}_3 &= x_2, & x_3(o) &= 0, \\ y &= x_3 + x_2x_1 . \end{aligned}$$

is minimal and has a Volterra series of length 3. It is in graded controllable canonical form with respect to the graded structure $\mathbb{R}^3 = (\mathbb{R} \oplus \mathbb{R}^2)$

$$\delta_t(x_1, x_2, x_3) = (tx_1, t^2x_2, t^2x_3)$$

In this case $n_1 = 1$, $n_2 = 2$, but $n_3 = 0$. This system is not in graded observable canonical form however. For future reference it is interesting to note that the observation space is $H_1 = \text{Span}\,\{x_3 + x_2x_1, x_2, x_1, x_1^2, x_1^3, 1\}$.

The same Volterra series may also be realized by the system

$$\begin{aligned} \dot{z}_1 &= u, & z_1(o) &= 0, \\ \dot{z}_2 &= z_1^2, & z_2(o) &= 0, \\ \dot{z}_3 &= z_2 + z_1^3 + z_2u, & z_3(o) &= 0, \\ y &= z_3, \end{aligned}$$

Here $z_3 = x_3 + x_2x_1$, $z_2 = x_2$, $z_1 = x_1$. This system is in graded observable canonical form with respect to the graded structure, $m_i = 1$, $1 \leq i \leq 3$, $\mathbb{R}^3 = \mathbb{R} \oplus \mathbb{R} \oplus \mathbb{R}$, $\delta_t(z_1,z_2,z_3) = (tz_1, t^2z_2, t^3z_3)$, but is not now in graded controllable canonical form. Note that in this case the observation space is

$$H_2 = \text{Span}\,\{z_3, z_2, z_1, z_1^2, z_1^3, 1\}.$$

3. FEEDBACK INVARIANT STRUCTURES

If we apply linear output feedback $u_i = \sum_j k_{ij}y_j + \bar{u}_i$ to a system (2) with Lie algebra L and observation space H we obtain a system

$$\begin{aligned} \dot{x} &= f(x) + \sum_{ij} k_{ij}g_i(x)h_j(x) + \sum_i \bar{u}_i g_i(x) \\ y_i &= h_i(x) \end{aligned} \qquad -6$$

In general the observation space H' and Lie algebra L' of this new system differ from H and L respectively. Let H_A denote the algebra generated by H under pointwise multiplication of functions. We then have

$$L' \subset H_A \otimes L, \quad H' \subset H_A$$

and in particular $H'_A \subset H_A$, $H'_A \otimes L' \subset H_A \otimes L$. However applying the feedback $\bar{u}_i = -\sum_j k_{ij} y_j + u_i$ to the system (6) we obtain

$$H_A \subset H'_A , \quad H_A \otimes L \subset H'_A \otimes L'$$

We conclude that H_A and $H_A \otimes L$ are invariant under feedback. We shall call H_A the observation algebra.

In contrast with linear systems, if we apply linear output feedback to a canonical realization of a s.f.v.s., we do not in general obtain a system in canonical form. This is one of the major drawbacks with this class of system. However the polynomial nature of the vector fields and output functions remains unaltered. In particular H_A and $H_A \otimes L$ will be algebras of polynomials and vector fields with polynomial coefficients respectively. In this section we give a description of these spaces for minimal realizations of s.f.v.s. in g.c.c.f. and g.o.c.f.

If $(\mathbb{R}^n, \delta_t)$ is a graded vector space of degree p, $\mathbb{R}^n = \mathbb{R}^{n_1} \oplus \dots \oplus \mathbb{R}^{n_p}$ denote by Q^k the polynomial algebra generated by 1 and the coordinate functions $x_1^1 \dots x_1^{n_1}\, x_2^1 \dots x_2^{n_2} \dots x_k^1 \dots x_k^{n_k}$ for $1 \leq k \leq p$. Q^o is the space of constant functions.

<u>LEMMA 1</u> For any realization of a s.f.v.s. in g.o.c.f. the observation algebra H_A is Q^p.

<u>PROOF</u> Since $H^{p-k} \subset C^k$ and $d\,H^{p-k}(o) = d\,C^k(o)$ for $1 \leq k \leq p$ a simple induction argument on k shows that H contains the functions

$$x_k^j + s_k^j (x_1 \dots x_{k-1}) , \quad 1 \leq j \leq n_k , \quad 1 \leq k \leq p$$

where s_k^j are polynomials. By theorem (2) H contains the constant functions, and in particular H contains the functions x_p^j, $1 \leq j \leq n_p$. Another simple induction argument shows that H_A contains all the coordinate functions x_k^j, $1 \leq j \leq n_k$, $1 \leq k \leq p$. In particular H_A coincides with Q^p.

We note that lemma (1) can in fact be trivial since as shown in Crouch and Collingwood [3] the g.o.c.f. guaranteed in theorem (6) may be chosen so that H contains all the coordinate functions. The example at the end of section (2) illustrates this point. The observation space H_1 for the realization in g.c.c.f. does not contain all of the coordinate functions, whereas the observation space H_2 for the realization in g.o.c.f. does contain all the coordinate functions. We also notice that the algebra generated by H_1 is equal to the polynomial algebra in all coordinate functions. This is a general result, although more difficult to prove.

THEOREM 7 CROUCH AND COLLINGWOOD [3]

In any minimal realization of a s.f.v.s. of length p in g.c.c.f. on a graded vector space ($\mathbb{R}^n$, δ_t) of degree p, the observation algebra H_A is equal to Q^p.

The following example shows that the observation space of a minimal system in g.c.c.f. under linear output feedback may in fact coincide with the polynomial algebra in the coordinate functions.

$$\dot{x} = u$$
$$\dot{z} = x^2$$
$$\cdot \quad y = z^3$$

Under the feedback $u = y + \bar{u}$ this system becomes

$$\dot{x} = z^3 + \bar{u}$$
$$\dot{x} = x^2$$
$$y = z^3$$

The Lie algebra of this system contains the vector fields

$$\frac{\partial}{\partial x}, \ \frac{\partial}{\partial z}, \ x\frac{\partial}{\partial z}, \ z\frac{\partial}{\partial x}, \ z^2\frac{\partial}{\partial x}$$

Clearly the observation space contains, 1, x, z, z^2, xz and x^2. Assuming by induction that the observation space contains $x^{n-p}z^p$ for $o \leq p \leq n$ application of $z^2\frac{\partial}{\partial x}$ to $x\,z^{n-1}$ shows that the observation space contains z^{n+1}. Repeated application of $x\frac{\partial}{\partial z}$ shows that the observation space contains $x^{n+1-p}z^p$ for $o \leq p \leq n+1$, completing the induction.

We now turn attention to the Lie algebra $H_A \otimes L$. The following result can be viewed as a dual to Lemma (1).

LEMMA 2 For any realization of a s.f.v.s. in g.c.c.f. on a graded vector space ($\mathbb{R}^n$, δ_t) of degree p the Lie algebra $Q^p \otimes L$ is the Lie algebra of all polynomial vector fields on $\mathbb{R}^n$.

PROOF $S^k \subset V^k$ and $S^k(o) = V^k(o)$ for $1 \leq k \leq p$ shows, by a simple induction argument on k, that S and hence L contains the vector fields

$$\partial/\partial x^j_k + \sum_{j>k} r^i_j(x_1 \ldots x_{j-1}) \ \partial/\partial x^j_k \quad 1 \leq j \leq n_k, \quad 1 \leq k \leq p.$$

where r^i_j are polynomials. In particular S contains $\partial/\partial x^j_k$ $1 \leq j \leq n_p$. Another simple induction argument shows that $Q^p \otimes L$ contains all the vector fields

$\partial/\partial x_p^j, 1 \leq j \leq n_k, \quad 1 \leq k \leq p$, and hence all polynomial vector fields.

This result may be generalised, in the same way in which lemma (1) was generalised by theorem (7). We give a full proof of this result since it does not appear elsewhere.

THEOREM 8

In any strongly accessible realization of a s.f.v.s. of length p in canonical form on a graded vector space $(\mathbb{R}^n, \delta_t)$ of degree p, the Lie algebra $Q^p \otimes L$ is the Lie algebra of all polynomial vector fields on $\mathbb{R}^n$.

PROOF We actually show that $Q^p \otimes S$ is the Lie algebra of all polynomial vector fields on $\mathbb{R}^n$. Set $K^i = S + (Q^{p-1} \otimes W^p \oplus Q^{p-2} \otimes W^{p-1} \oplus \ldots \oplus Q^i \otimes W^{i+1} \oplus Q^{i-1} \otimes W^i)$ with $K^{p+1} = S$. Note that if $h \in Q^j$ and $X \in Q^{k-1} \otimes W^k$ then $L_X(h) \in Q^j$ if $k \leq j$ and $L_X(h)$ is the zero function if $j < k$. Since

$$S \subset V^1 \subset \bigoplus_{k=1}^{p} (C^{k-1} \otimes W^k) \subset \bigoplus_{k=1}^{i-1} (Q^{k-1} \otimes W^k) \oplus \bigoplus_{k=i}^{p} (Q^{k-1} \otimes W^k)$$

we see that each linear space K^i is a Lie algebra. Moreover each of the linear spaces $Q^j \otimes K^i$ is also a Lie algebra.

The main step in the proof is to show by induction on i that $W^i \subset Q^{i-1} \otimes K^{i+1}$ for $1 \leq i \leq p$. To show that this is true for $i = 1$ we show $W^1 \subset K^2$. However $S^2(x) \subset V^2(x) \subset W^p(x) + \ldots + W^2(x)$ for all $x \in \mathbb{R}^n$, so by strong accessibility there exist vector fields in S, and not in S^2 whose span contains $W^1(x)$. It follows that S contains the vector fields

$$\partial/\partial x_1^j + \sum_{\substack{i>1 \\ k}} r_i^k(x_1 \ldots x_{i-1}) \, \partial/\partial x_i^k, \quad 1 \leq j \leq n_1$$

where r_i^k are polynomials. Since $S \subset K^2$ it follows that K^2 contains W^1.

Now assume that $W^r \subset Q^{r-1} \otimes K^{r+1}$ for all $r < i$. We shall prove $W^i \subset Q^{i-1} \otimes K^{i+1}$. We claim that for each $r < i$ we have

$$W^r \subset \bigoplus_{j=r}^{i-1} Q^j \otimes W^{j+1} + Q^{r-1} \otimes K^{i+1} \qquad -7$$

Now $K^{r+1} = K^{i+1} + \bigoplus_{j=r}^{i-1} Q^j \otimes W^{j+1}$ and by induction $W^r \subset Q^{r-1} \otimes K^{r+1}$ so

$$W^r \subset \bigoplus_{j=r}^{i-1} Q^j \otimes W^{j+1} + Q^{r-1} \otimes K^{i+1}.$$

We now show by induction on $s \leq i$ that for each j, $1 \leq j \leq n_i$, $Q^{s-2} \otimes K^{i+1}$ contains the vector fields

$$\sum_{\substack{\ell=s\\k}}^{i} \alpha_\ell^k(x_1 \ldots x_{\ell-1})\,\partial/\partial x_\ell^k \qquad -8$$

where α_ℓ^k are polynomials satisfying

$$\alpha_\ell^k\,(0 \ldots 0) = 0 \quad \text{for} \quad \ell \neq i \qquad k$$

$$\alpha_i^k\,(0 \ldots 0) = 0 \quad \text{for} \quad k \neq j \qquad -9$$

$$\alpha_i^j\,(0 \ldots 0) \neq 0$$

To prove this for s = 1 we notice that strong accessibility implies that *S* contains vector fields for each j, $1 \leq j \leq n_i$ given by

$$\sum_{\substack{\ell=1\\k}}^{p} \beta_\ell^k(x_1 \ldots x_{\ell-1})\ \partial/\partial x_\ell^k$$

where β_ℓ^k are polynomials satisfying (9). It follows that K^{i+1} contains the required vector fields satisfying (9), since it contains $\bigoplus_{\ell=p}^{i+1} Q^{\ell-1} \otimes W^\ell$. Setting $Q^{-1} = Q^0$ we have proved the assertion for s = 1. Now assume that the assertion is true for s with $s \leq r \leq i$. By (7) for each j, $1 \leq j \leq n_r$ there exist polynomials γ_ℓ^{jk} such that the vector fields $\partial/\partial x_r^j + \sum_{\ell=r+1}^{i} \gamma_\ell^{jk}(x_1 \ldots x_{\ell-1})\,\partial/\partial x_\ell^k$ belong to $Q^{r-1} \otimes K^{i+1}$. Multiply these vector fields by the polynomials $\alpha_r^j(x_1 \ldots x_{r-1}) \in Q^{r-1}$ in (9) to obtain vector fields

$$\alpha_r^j(x_1 \ldots x_{r-1})\,\partial/\partial x_r^i + \sum_{\ell=r+1}^{i} \bar{\gamma}_\ell^{jk}\ (x_1 \ldots x_{\ell-1})\,\partial/\partial x_\ell^k$$

which belong to $Q^{r-1} \otimes K^{i+1}$ and satisfying $\bar{\gamma}_\ell^{jk}\ (0 \ldots 0) = 0$.

Subtracting these vector fields for $1 \leq j \leq n_r$ from that in (8) with s=r, shows that Q^{r-1} O K^{i+1} contains for each j, $1 \leq j \leq n_i$, the vector fields

$$\sum_{\ell=r+1}^{i} \bar{\alpha}_\ell^k\ (x_1 \ldots x_{\ell-1})\,\partial/\partial x_\ell^k$$

where $\bar{\alpha}_\ell^k$ are polynomials satisfying the conditions (9). We have therefore proved the assertion for s = r + 1, thereby completing the induction.

We conclude that $Q^{i-2} \otimes K^{i+1}$ contains for each j, $1 \leq j \leq n_i$ the vector fields

$$Y_1^j(x) = \sum_k a^k(x_1 \ldots x_{i-1})\ \partial/\partial x_i^k \in Q^{i-1} \otimes W^i$$

where a^k are polynomials satisfying

$$a^k(0 \dots 0) = 0 \quad \text{for } k \neq j$$

$$a^k(0 \dots 0) = 1 \quad \text{for } k = j$$

We deduce from (7) with $r = i - 1$ that there exist polynomials β_i^k such that the vector fields

$$X_{i-1}^{\ell}(x) = \partial/\partial x_{i-1}^{\ell} + \sum_k \beta_i^k (x_1 \dots x_{i-1})\, \partial/\partial x_i^k$$

belong to $Q^{i-2} \otimes K^{i+1}$ for each ℓ, $1 \leq \ell \leq n_{i-1}$. Now since $Q^{i-2} \otimes K^{i+1}$ is a Lie algebra we have $ad^k(X_{i-1}^{\ell})Y_1^j = ad^k(\partial/\partial x_{i-1}^{\ell})(Y_1^j) \in Q^{i-2} \otimes K^{i+1}$

for each $k \geq 0$, where $ad^k(X)(Y) = adX(ad^{k-1}(X)(Y))$ and $adX(Y) = [X,Y] = ad^1(X)(Y)$. Identifying Y_1^j as a vector in $\mathbb{R}^{n_i}$ with polynomial components a^k we may expand Y_1^j to obtain

$$Y_1^j(x) = \sum_{s=0}^{N} (x_{i-1}^1)^s \, a_s(x_1 \dots x_{i-2}, \hat{x}_{i-1})$$

where a_s are vectors in $\mathbb{R}^{n_i}$ with polynomial components in all components of $x_1 \dots x_{i-1}$ except x_{i-1}^1. Lie differentiation of Y_1^j by X_{i-1}^1 now shows that a_N can be identified with a vector field in $Q^{i-2} \otimes K^{i+1}$. It follows that

$$\sum_{s=0}^{N-1} (x_{i-1}^1)^s \, a_s(x_1 \dots x_{i-2}, \hat{x}_{i-1})$$

also belongs to $Q^{i-1} \otimes K^{i+1}$. Repeating this process shows that a_0 can be identified with a vector field in $Q^{i-1} \otimes K^{i+1}$. Note that $a_0(0 \dots 0)$ is identified with $\partial/\partial x_i^j$.

Repeated expansions about $x_{i-1}^2 \dots x_{i-1}^{n_{i-1}}$ in turn and Lie differentiation with $X_{i-1}^2 \dots X_{i-1}^{n_{i-1}}$ show that there exists a vector field $Y_2^j(x_1 \dots x_{i-2})$ belonging to $Q^{i-1} \otimes K^{i+1}$ such that $Y_2^j(0 \dots 0) = \partial/\partial x_i^j$, and $Y_2^j \in Q^{i-2} \otimes W^i$.

We again deduce from (7) with $r = i-2$ that there exists polynomials β_{i-1}^k and β_i^k such that the vector fields

$$X_{i-2}^{\ell} = \partial/\partial x_{i-2}^{\ell} + \sum_k \beta_{i-1}(x_1 \dots x_{i-2})\, \partial/\partial x_{i-1}^k + \sum_k \beta_i^k(x_1 \dots x_{i-1})\, \partial/\partial x_i^k$$

belong to $Q^{i-3} \otimes K^{i+1}$ for each ℓ, $1 \leq \ell \leq n_{i-2}$. Since $Q^{i-3} \otimes K^{i+1} \subset Q^{i-1} \otimes K^{i+1}$ which is a Lie algebra we have

$$ad^k(X_{i-2}^{\ell})(Y_2^j) = ad^k(\partial/\partial x_{i-2}^{\ell})(Y_2^j) \in Q^{i-1} \otimes K^{i+1}.$$

In particular we may eliminate the x^{ℓ}_{i-2} coordinates from Y^j_2 as before to obtain a vector field $Y^j_3(x_1 \dots x_{i-3}) \in Q^{i-1} \otimes K^{i+1}$ which satisfies $Y^j_3(0 \dots 0) = \partial/\partial x^j_i$ and $Y^j_3 \in Q^{i-3} \otimes W^i$.

A simple induction argument using (7) now shows that $Q^{i-1} \otimes K^{i+1}$ contains a constant vector field $Y^j_i = \partial/\partial x^j_i$. Repeating this argument for each j. $1 \leq j \leq n_i$ shows that $W^i \subset Q^{i-1} \otimes K^{i+1}$, which completes the induction on i.

In particular $W^p \subset Q^{p-1} \otimes K^{p+1} = Q^{p-1} \otimes S$, so that $K^p = Q^{p-1} \otimes S$. However for $p > r \geq 1$ $Q^{p-1} \otimes K^r = Q^{p-1} \otimes (K^{r+1} + Q^{r-1} \otimes W^r) = Q^{p-1} \otimes K^{r+1} + Q^{p-1} \otimes W^r$. By the above result $Q^{p-1} \otimes W^r \subset Q^{p-1} \otimes K^{r+1}$ so $Q^{p-1} \otimes K^r = Q^{p-1} \otimes K^{r+1}$. Therefore $Q^{p-1} \otimes K^1 = Q^{p-1} \otimes S$ and in particular $Q^p \otimes S = Q^p \otimes (W^1 \oplus \dots \oplus W^p)$ which is the Lie algebra of all polynomial vector fields which completes the proof of the theorem.

With this result we may give the desired conclusion of this section concerning $H_A \otimes L$.

THEOREM 9

In any minimal realization of a s.f.v.s. of length p, on a graded vector space $(\mathbb{R}^n, \delta_t)$ of degree p, in either g.c.c.f. or g.o.c.f. the Lie algebra $H_A \otimes L$ is the Lie algebra of all polynomial vector fields.

PROOF If the realization is in g.c.c.f. then by theorem (7) $H_A = Q^p$ and by lemma (2) $H_A \otimes L$ is the lie algebra of all polynomial vector fields. If the realization is in g.o.c.f. then by lemma (1) $H_A = Q^p$, and by theorem (8) the Lie algebra $H_A \otimes L$ is the Lie algebra of all polynomial vector fields.

We remark that we do not require both theorems (8) and (7) to prove this result in either of the cases given above. Note also that the results of this section are all trivially satisfied for minimal linear systems. Thus although the class of canonical realizations of s.f.v.s. of length greater than one is not closed under linear output feedback, the feedback invariant spaces H_A and $H_A \otimes L$ have exactly the same characterization for realization of s.f.v.s. of length greater than one and equal to one. For further applications of these spaces in the filtering theory for realizations of s.f.v.s., see Crouch and Collingwood [3].

For our final observation we consider so called Hamiltonian systems of the form

$$\dot{q} = \frac{\partial H_o}{\partial p}(q,p)' + u\,\frac{\partial H_1}{\partial p}(q,p)' , \qquad q \in \mathbb{R}^n$$

$$-\dot{p} = \frac{\partial H_o}{\partial q}(q,p)' + u\,\frac{\partial H_1}{\partial q}(q,p)' , \qquad p \in \mathbb{R}^n$$

$$y = H_1(q,p)$$

For more details concerning such systems see Van der Schaft [12] and references contained therein. We note that this class of system is closed under linear feedback $u = k\,H_1(q,p) + \bar{u}$ since we simply replace the Hamiltonian $H_o(q,p)$ by

$$H_o + k/2\; H_1(q,p)^2 .$$

As demonstrated in Crouch and Irving [4] we have no difficulty in finding Hamiltonian systems which are simultaneously in g.c.c.f. and g.o.c.f., and have input output maps given by s.f.v.s. The system

$$\dot{q} = u\,, \quad -\dot{p} = q^2\,, \quad y = p$$

is a simple trivial example as observed earlier.

For such systems it is however clear that the Lie algebra of the system under linear feedback can never equal the Lie algebra of all polynomial vector fields; since not all polynomial vector fields can be expressed as Hamiltonian vector fields, with respect to a given symplectic structure. It follows from theorem (9) that the Lie algebra under feedback for such systems will never attain $H_A \otimes L$.

REFERENCES

1. R.W. BROCKETT: "Volterra Series and Geometric Control Theory", Automatica, Vol. 12, pp. 167-176 (1976).

2. P.E. CROUCH: "Dynamical Realizations of Finite Volterra Series", S.I.A.M.J. of Control and Optimization, Vol. 19, pp. 177-202 (1981).

3. P.E. CROUCH & P.C. COLLINGWOOD: "The Observation Space and Realizations of Finite Volterra Series", Control Theory Centre Report No. 117, University of Warwick, Coventry (1983).

4. P.E. CROUCH & M. IRVING: "Dynamical Realizations of Homogeneous Hamiltonian Systems", Control Theory Centre Report No. 122, University of Warwick, Coventry (1984).

5. M. FLIESS & I. KUPKA: "A Finiteness Criterion for Nonlinear Input-output Differential Systems". To appear in S.I.A.M. J. of Control and Optimization (1983).

6. R.W. GOODMAN: "Nilpotent Lie Groups: Structure and Applications to Analysis", Sprinter Verlag, Lecture Notes in Mathematics No. 562 (1976).

7. J.P. GAUTHIER & G. BORNARD: "Observability for any u(t) of a Class of Nonlinear Systems", I.E.E.E. Trans., Vol. A.C. 26, pp. 922-926 (1981).

8. R. HERMANN & A.J. KRENER: "Nonlinear Observability and Controllability", I.E.E.E. Trans., Vol. AC 23, pp. 1090-1095 (1978).

9. H. NIJMEIJER: "Observability of a Class of Nonlinear Systems. A Geometric Approach", Ricerche Di Automatica, Vol. 12, pp. 50-68 (1981).

10. H.J. SUSSMANN: "Existence and Uniqueness of Minimal Realizations of Nonlinear Systems", Math. Systems Theory, pp. 263-284, Vol. 10 (1977).

11. H.J. SUSSMANN & V. JURDEVIC: "Controllability of Nonlinear Systems", Journal of Differential Equations, Vol. 12, pp. 313-329 (1972).

12. A.J. VAN DER SCHAFT: "Controllability and Observability for Affine Nonlinear Systems", I.E.E.E. Trans., Vol. AC-27, pp. 490-492 (1982).

ON THE ORDER REDUCTION OF LINEAR OPTIMAL CONTROL SYSTEMS IN CRITICAL CASES

A.L. Dontchev and V.M. Veliov
Institute of Mathematics
Bulgarian Academy of Sciences
1090 Sofia, P.O. Box 373
Bulgaria

Abstract.
Linear control systems with a small parameter in the derivatives and constrained controls are considered. On the assumption that the fundamental solution of the fast subsystem is bounded (but not necessarily asymptotically stable) the behaviour of the set of trajectories is investigated when the small parameter tends to zero. For the periodic case the Hausdorff limit of the reachable set is derived. The well-posedness of the order reduction of various optimal control systems is justified.

1. Introduction.

This paper deals with singularly perturbed linear control systems

$$(1)\qquad \begin{aligned} \dot{x} &= A_1 x + A_2 y + B_1 u, \\ \varepsilon\dot{y} &= A_3 x + A_4 y + B_2 u, \end{aligned}$$

where $x(t) \in R^n$ is the slow state, $y(t) \in R^m$ is the fast state, the positive scalar ε represents the singular perturbation, the control u takes values from given compact constraining set

$$u(t) \in U \subset R^r.$$

For $\varepsilon = 0$ the dimension of the state space of (1) reduces from n+m to n since the differential equation for y becomes an algebraic equation

$$0 = A_3 x + A_4 y + B_2 u.$$

By solving this equation one can eliminate y from the first equation obtaining the reduced order system. Clearly, this procedure may lead to a significant simplification of the original model (1).

Recently, the order reduction of control systems has been studied in a number of papers, for a survey see e.g. Kokotovic [1].

In the works [2] and [3] we study the well-posedness of the order reduction in constrained optimal control problems under various hypotheses for the state matrix A_4. In our analysis the continuity of the set values map: "parameter $\varepsilon \longrightarrow$ set of trajectories of (1)" plays a crucial role. Here the so-called critical case is considered when the eigenvalues of the matrix A_4 may have zero real parts. In contrast to the paper [4] we do not confine the analysis to the slow part of the trajectories; however, we assume that $B_2 = 0$.

In Section 2 we give two basic lemmas concerning the limit properties of the trajectories of (1) as $\varepsilon \longrightarrow 0$. Section 3 shows that the reachable set is lower semicontinuous at $\varepsilon = 0$. This follows from a general result on the continuity of integrals of set valued maps. As applications, in Sections 4 and 5 we study the behaviour of the optimal value (marginal function) under a change of the system order. In Section 6 we formulate two open problems, which, in our opinion, are important for the further investigations.

In order to simplify the presentation we consider time-invariant systems only; the mathematical techniques, however, can be easily extended to linear systems with time-varying matrices and time-depending control constraining set under suitable continuity hypotheses.

Throughout the paper $|\cdot|$ is the euclidean norm while $\|\cdot\|_2$ is the L_2 norm and $\|\cdot\|_c$ is the uniform norm. By $\delta(\varepsilon)$ we denote a (vector) function which tends to zero with ε (uniformly in the time t).

2. Convergence of the trajectories.

Consider the following singularly perturbed system

$$\dot{x} = A_1 x + A_2 y + Bu \quad , \; x(0) = x^0$$

$$(2) \quad \varepsilon\dot{y} = A_3 x + (A_4 + \varepsilon A_4')y \quad , \; y(0) = y^0$$

$$x(t) \in R^n, \; y(t) \in R^m, \; t \in [0,T]$$

with admissible set of control functions

$$u(\cdot) \in \mathcal{U} = \{u(\cdot)\text{-measurable, } u(t) \in U \text{ for a.e. } t \in [0,T]\}.$$

We assume that:

(A1) The matrix A_4 is invertible and $\limsup_{t\to+\infty} |\exp(A_4 t)| < +\infty$.
The set U is compact.

The reduced system has the form

$$(3) \quad \begin{aligned} \dot{x} &= A_o x + Bu, \quad x(0) = x^o, \\ y &= -A_4^{-1} A_3 x, \end{aligned}$$

where

$$A_o = A_1 - A_2 A_4^{-1} A_3.$$

Lemma 1.

There exist $\varepsilon_0 > 0$ and a constant C such that if $\varepsilon \in (0, \varepsilon_0]$ and the control $u_\varepsilon(\cdot) \in \mathcal{U}$ are arbitrarily chosen, and $(x_\varepsilon(\cdot), y_\varepsilon(\cdot))$ solves (2) with $u = u_\varepsilon$ and ε, then

$$\|x_\varepsilon\|_C + \|y_\varepsilon\|_C < C.$$

Moreover, for all $t > 0$

$$(4) \quad x_\varepsilon(t) = \exp(A_o t) + \int_0^t \exp(A_o(t-s)) B u_\varepsilon(s) ds + \delta(\varepsilon),$$

$$(5) \quad y_\varepsilon(t) = \exp(A_4 t/\varepsilon) \exp(A_4' t)(y^o + A_4^{-1} A_3 x^o) - A_4^{-1} A_3 \exp(A_o t) x^o + \\ + \int_0^t (A_4^{-1} A_3 \exp(A_o(t-s)) + \exp((A_4 + \varepsilon A_4')(t-s)/\varepsilon) A_4^{-1} A_3) B u_\varepsilon(s) ds + \delta(\varepsilon).$$

Proof. Using (A1) and integrating by parts we obtain

$$(6) \quad \begin{aligned} x_\varepsilon(t) &= x^o + \int_0^t A_1 x_\varepsilon(s) ds + \int_0^t A_2 \exp((A_4 + \varepsilon A_4') s/\varepsilon) y^o ds \\ &+ \frac{1}{\varepsilon} \int_0^t A_2 \int_0^s \exp((A_4 + \varepsilon A_4')(s-\tau)/\varepsilon) A_3 x_\varepsilon(\tau) d\tau ds + \int_0^t B u_\varepsilon(s) ds \\ &= x^o + \int_0^t A_1 x_\varepsilon(s) ds + \frac{1}{\varepsilon} \int_0^t \int_\tau^t A_2 \exp((A_4 + \varepsilon A_4')(s-\tau)/\varepsilon) A_3 x_\varepsilon(\tau) ds d\tau \end{aligned}$$

$$+ \int_0^t Bu_\varepsilon(s)ds + \delta(\varepsilon)$$

$$= x^o + \int_0^t A_1 x_\varepsilon(s)ds + \int_0^t\int_\tau^t A_2 \frac{d}{ds}\exp((A_4+\varepsilon A_4')(s-\tau)/\varepsilon)(A_4+\varepsilon A_4')^{-1}\times$$

$$A_3 x_\varepsilon(\tau)dsd\tau + \int_0^t Bu_\varepsilon(s)ds + \delta(\varepsilon)$$

(6)

$$= x^o + \int_0^t A_1 x_\varepsilon(s)ds + \int_0^t A_2\exp((A_4+\varepsilon A_4')(t-\tau)/\varepsilon)(A_4+\varepsilon A_4')^{-1}A_3 x_\varepsilon(\tau)d\tau$$

$$- \int_0^t A_2(A_4+\varepsilon A_4')^{-1}A_3 x_\varepsilon(s)ds + \int_0^t Bu_\varepsilon(s)ds + \delta(\varepsilon).$$

The compactness of U and Gronwall lemma yield uniform boundedness of $x_\varepsilon(\cdot)$.
Furthermore

$$y_\varepsilon(t) = \exp((A_4+\varepsilon A_4')t/\varepsilon)y^o + \frac{1}{\varepsilon}\int_0^t \exp((A_4+\varepsilon A_4')(t-s)/\varepsilon)A_3 x_\varepsilon(s)ds$$

$$= \exp((A_4+\varepsilon A_4')t/\varepsilon)y^o - \int_0^t \frac{d}{ds}\exp((A_4+\varepsilon A_4')(t-s)/\varepsilon)(A_4+\varepsilon A_4')^{-1}A_3 x_\varepsilon(s)ds$$

$$= \exp((A_4+\varepsilon A_4')t/\varepsilon)y^o - (A_4+\varepsilon A_4')^{-1}A_3 x_\varepsilon(t) +$$

(7)

$$+ \exp((A_4+\varepsilon A_4')t/\varepsilon)(A_4+\varepsilon A_4')^{-1}A_3 x^o$$

$$+ \int_0^t \exp((A_4+\varepsilon A_4')(t-s)/\varepsilon)(A_4+\varepsilon A_4')^{-1}A_3(A_1 x_\varepsilon(s) +$$

$$+ A_2 y_\varepsilon(s) + Bu_\varepsilon(s))ds.$$

Applying again Gronwall lemma we get that $y_\varepsilon(\cdot)$ is uniformly bounded. Thus, $\dot{x}_\varepsilon(\cdot)$ is uniformly bounded and

$$\int_0^t A_2\exp((A_4+\varepsilon A_4')(t-s)/\varepsilon)(A_4+\varepsilon A_4')^{-1}A_3 x_\varepsilon(\tau)d\tau = \delta(\varepsilon).$$

This, combined with (6) gives us (4). From (7) we have

$$y_\varepsilon(t) = \exp((A_4+\varepsilon A_4')t/\varepsilon)(y^o + A_4^{-1}A_3 x^o) - A_4^{-1}A_3 x_\varepsilon(t)$$

$$+ \int_0^t \exp((A_4+\varepsilon A_4')(t-s)/\varepsilon)(A_4+\varepsilon A_4')^{-1}A_3(A_o x_\varepsilon(s) + Bu_\varepsilon(s))ds + \delta(\varepsilon).$$

Taking into account (4) and the boundedness of $\|\dot{x}_\varepsilon\|_C$ we obtain (5). □

Lemma 2.

Suppose that the feasible control $u_\varepsilon(\cdot)$ tends L_2-weakly to $u_0(\cdot)$ as $\varepsilon \to 0$. Let $(x_\varepsilon(\cdot), y_\varepsilon(\cdot))$ be the corresponding solution of (2) and $(x_0(\cdot), y_0(\cdot))$ solve (3) for $u_0(\cdot)$.
Then

$$x_\varepsilon(\cdot) \to x_0(\cdot) \text{ in } C^{(n)}[0,T],$$
$$y_\varepsilon(\cdot) \to y_0(\cdot) \text{ weakly in } L_2^{(m)}[0,T]$$

as $\varepsilon \to 0$. If, in addition, $u_\varepsilon(\cdot) = u_0(\cdot)$ for $\varepsilon > 0$ and

$$y^o + A_4^{-1}A_3x^o = 0,$$

then

$$y_\varepsilon(\cdot) \to y_0(\cdot) \text{ in } C^{(m)}[0,T] \text{ as } \varepsilon \to 0.$$

The proof is entirely based on (4), (5) and the relations

$$\lim_{\varepsilon\to 0} \int_0^\tau\int_0^t \exp((A_4+\varepsilon A_4')(t-s)/\varepsilon)A_4^{-1}A_3Bu_\varepsilon(s)dsdt = 0$$

for any fixed $\tau \in [0,T]$ and

$$\max_{0<t<T} \int_0^t \exp((A_4+\varepsilon A_4')(t-s)/\varepsilon)A_4^{-1}A_3Bu_0(s)ds = \delta(\varepsilon).$$ □

Remark 1.

The above lemmas do not change if one replaces the initial condition (x^o,y^o) by $(x_\varepsilon^o,y_\varepsilon^o)$ such that $x_\varepsilon^o \to x^o$ and $y_\varepsilon^o \to y^o$ when $\varepsilon \to 0$.

3. Convergence of the reachable set.

Denote as $G(T,\varepsilon)$ the reachable set at some fixed $T > 0$ of the system (2), that is

$$G(T,\varepsilon) = \{(x,y)\ R^{n+m}, (x,y)=(x_\varepsilon(T),y_\varepsilon(T)) \text{ where } (x_\varepsilon(\cdot),y_\varepsilon(\cdot)) \text{ solves (2) for some } u_\varepsilon(\cdot) \in \mathcal{U}\}.$$

In this section we study the continuity properties of $G(T,\cdot)$ at $\varepsilon = 0$ assuming that (A1) and the following condition hold:

(A2) The matrix function $\exp(A_4 t)$ is periodic with period $\omega > 0$ and $y^o + A_4^{-1}A_3x^o = 0$.

Define the compact and convex sets

$$R(t) = \frac{1}{\omega}\int_0^{\omega}\begin{bmatrix}\exp(A_o t)\\ -A_4^{-1}A_3\exp(A_o t)+\exp(A_4 s)\exp(A_4' t)A_4^{-1}A_3\end{bmatrix} BUds$$

and

$$(8)\ G(T) = \begin{bmatrix}\exp(A_o T)x^o\\ -A_4^{-1}A_3\exp(A_o T)x^o\end{bmatrix} + \int_0^T R(t)dt.$$

Lemma 3.

For every $T > 0$

$$h(G(T,\varepsilon),G(T)) \longrightarrow 0 \text{ as } \varepsilon \longrightarrow 0,$$

where $h(\cdot,\cdot)$ denotes the Hausdorff distance.

This lemma follows from the relations (4) and (5) in Lemma 1 and the following general result:

Theorem 1.

Let $F(t,s)$ be a compact valued and Hausdorff continuous map from R^2 to R^p and let $F(t,s+\omega) = F(t,s)$ for some $\omega > 0$ and for all t and s. For fixed $T > 0$ denote

$$W_\varepsilon = \int_0^T F(t,\tfrac{T-t}{\varepsilon})dt$$

and

$$W_o = \frac{1}{\omega}\int_0^T\int_0^{\omega} F(t,s)dsdt$$

$$= \{\int_0^T f(t)dt,\ f(t) \in \frac{1}{\omega}\int_0^{\omega} F(t,s)ds, f(\cdot) - \text{measurable}\}.$$

Then

$$h(W_\varepsilon,W_o) \longrightarrow 0 \text{ as } \varepsilon \longrightarrow 0.$$

In the proof we use:

Proposition 1. Let $G_1(t)$ and $G_2(t)$ be two continuous and compact valued maps from R^1 to R^p and

$$\max_{t\in[0,T]} h(G_1(t),G_2(t)) \leqslant \alpha.$$

Then for any measurable selection $g_1(\cdot)$ of $G_1(\cdot)$ there exists a measurable selection $g_2(\cdot)$ of $G_2(\cdot)$ such that

$$\operatorname{vraisup}_{t\in[0,T]} |g_1(t)-g_2(t)| \leqslant \alpha. \qquad \square$$

Proof of Theorem 1. Denote

$$Q(t) = \frac{1}{\omega}\int_0^{\omega} F(t,s)ds.$$

The continuous set valued map Q is compact and convex valued and

$$W_o = \int_0^T Q(t)dt.$$

First, let $w_\varepsilon \in W_\varepsilon$. Then there exists a measurable selection $f_\varepsilon(\cdot)$, $f_\varepsilon(t) \in F(t,\frac{T-t}{\varepsilon})$ such that

$$w_\varepsilon = \int_0^T f_\varepsilon(t)dt = \sum_{k=0}^{[\frac{T}{\omega\varepsilon}]-1} \int_{T-(k+1)\omega\varepsilon}^{T-k\omega\varepsilon} f_\varepsilon(t)dt + \delta(\varepsilon)$$

$$= \varepsilon \sum_{k=0}^{[\frac{T}{\omega\varepsilon}]-1} \int_0^{\omega} f_\varepsilon(T-k\omega\varepsilon-\varepsilon s)ds + \delta(\varepsilon)$$

$$= \varepsilon \sum_{k=0}^{[\frac{T}{\omega\varepsilon}]-1} \int_0^{\omega} g_\varepsilon^k(s)ds + \delta(\varepsilon),$$

where

$$g_\varepsilon^k(s) \in F(T-k\omega\varepsilon-\varepsilon s,s+k\omega) = F(T-k\omega\varepsilon-\varepsilon s,s)$$

$$\subset F(T-k\omega\varepsilon,s) + \delta(\varepsilon)B_p,$$

(B_p is the unique ball in R^p). Denote

$$\alpha_k^\varepsilon = \frac{1}{\omega}\int_0^T g_\varepsilon^k(s)ds.$$

Thus, (Proposition 1)

$$\alpha_k^\varepsilon \in Q(T-k\omega\varepsilon),\ k=0,\dots,[\frac{T}{\omega\varepsilon}]-1$$

and we have

$$w_\varepsilon = \varepsilon\omega \sum_{k=0}^{[\frac{T}{\omega\varepsilon}]-1} \alpha_k^\varepsilon + \delta(\varepsilon) = \sum_{k=0}^{[\frac{T}{\omega\varepsilon}]-1} \int_{k\omega\varepsilon}^{(k+1)\omega\varepsilon} \alpha_k^\varepsilon dt + \delta(\varepsilon)$$

$$= \int_0^T q_\varepsilon(t)dt + \delta(\varepsilon),$$

where $q_\varepsilon(t) = \alpha_k^\varepsilon$ for $t \in [k\omega\varepsilon,(k+1)\omega\varepsilon)$, $k=0,\dots[\frac{T}{\omega\varepsilon}]-1$.

Furthermore, if $\bar{q}_\varepsilon(T-t) \in Q(k\omega\varepsilon)$, $t \in [k\omega\varepsilon,(k+1)\omega\varepsilon)$ then

$$w_\varepsilon = \int_0^T \bar{q}_\varepsilon(t)dt + \delta(\varepsilon).$$

From Proposition 1 it follows that there exists a measurable function $\tilde{q}_\varepsilon(\cdot)$, $\tilde{q}_\varepsilon(t) \in Q(t)$ such that $\underset{t\in[0,T]}{\text{vraisup}} |\tilde{q}_\varepsilon(t)-\bar{q}(t)| = \delta(\varepsilon)$, thus

$$w_\varepsilon = \int_0^T \tilde{q}_\varepsilon(t)dt + \delta(\varepsilon) \in \int_0^T Q(t)dt + \delta(\varepsilon)B_p.$$

Hence, if

$$w_\varepsilon \longrightarrow w_o \text{ as } \varepsilon \longrightarrow 0$$

then

$$w_o \in W_o.$$

By repeating the above argument in the reverse way one can show that for every $w_o \in W_o$ there exists $w_\varepsilon \in W_\varepsilon$ such that $w_\varepsilon \longrightarrow w_o$ as $\varepsilon \longrightarrow 0$. The proof is complete. □

Remark 2.

The reachable set for the low-order system (3) is $G(T,o) = \{(x,y), x\in G_x(T,o), y = -A_4^{-1}A_3x\}$, where $G_x(T,0)$ is the reachable set of the reduced system

$$\dot{x} = A_o x + Bu, \; x(0) = x^o$$
$$u(t) \in U.$$

Denote by $G_x(T,\varepsilon)$ the projection of $G(T,\varepsilon)$ on R^n, i.e. the x-part of $G(T,\varepsilon)$. Lemma 3 implies that $G_x(T,\varepsilon)$ is Hausdorff continuous (from the left) at $\varepsilon = 0$. Nevertheless, this is not true for the entire set $G(T,\varepsilon)$ since

$$G(T) \supsetneq G(T,0),$$

that is the reachable set $G(T,\varepsilon)$ is lower semi-continuous at $\varepsilon = 0$, see the following example:

Example 1.

$$\begin{aligned} \dot{x} &= u, & x(0) &= 0, \quad u(t) \in U = [-1,1], \\ \varepsilon\dot{y}_1 &= y_2 + x, & y_1(0) &= 0, \\ \varepsilon\dot{y}_2 &= -y_1, & y_2(0) &= 0. \end{aligned}$$

The reduced system is

$$\begin{aligned} \dot{x} &= u, \quad x(0) = 0, \\ y_2 &= -x, \\ y_1 &= 0, \end{aligned}$$

hence

$$G(T,0) = \{(x,y_1,y_2), x\in[-1,1], \; y_1=0, \; y_2=-x\}.$$

We have

$$R(t) = R = \{\frac{1}{2\pi} \int_0^{2\pi} \begin{bmatrix} 1 \\ \cos s \\ -1-\sin s \end{bmatrix} u(s)ds, \; u(s)\in[-1,1]\}$$

and

$$G(1) = R.$$

For the feasible control $u(s) = \cos s$ we get the point $(0,2,0)$ which if from $G(1)$ but not from $G(1,0)$.

4. Well-posedness of the order reduction for optimal control systems.

Let $J(\cdot)$ be a functional defined for every feasible control $u(\cdot) \in \mathcal{U}$ and for the corresponding trajectory $(x_\varepsilon(\cdot), y_\varepsilon(\cdot))$ of (2). For fixed $\varepsilon > 0$ consider the optimal control problem

$$(P_\varepsilon): \quad \begin{array}{l} \text{Minimize } J(u(\cdot),x(\cdot),y(\cdot)) \\ \text{subject to } u(\cdot) \in \mathcal{U} \text{ and (2).} \end{array}$$

For $\varepsilon = 0$ we define the reduced problem

$$(P_0): \quad \begin{array}{l} \text{Minimize } J(u(\cdot),x(\cdot),y(\cdot)) \\ \text{subject to } u(\cdot) \in \mathcal{U} \text{ and (3).} \end{array}$$

The following theorem (without proof) is based on Lemma 2:

Theorem 2.

Suppose that (A1) holds and the functional $J(\cdot)$ is lower semi-continuous with respect to $(u(\cdot),y(\cdot))$ in the weak L_2 topology and with respect to $x(\cdot)$ in the uniform topology. Let one of the following conditions hold:

(i) for $u(\cdot)$ fixed $J(u(\cdot),\cdot,\cdot)$ is linear with respect to $y(\cdot)$ and continuous with respect to $x(\cdot)$ in the uniform topology.

(ii) $y^0 + A_4^{-1}A_3x^0 = 0$ and for $u(\cdot)$ fixed $J(u(\cdot),\cdot,\cdot)$ is continuous in the uniform topology.

Then the problems (P_ε) and (P_0) have solutions (for sufficiently small $\varepsilon>0$) and if $\hat{J}_\varepsilon$ and $\hat{J}_0$ are the optimal values then

$$\hat{J}_\varepsilon \longrightarrow \hat{J}_0 \text{ as } \varepsilon \longrightarrow 0.$$

Hence, as far as the optimal value of (P_0) approximates the optimal value of (P_ε) for small $\varepsilon > 0$ the order reduction is well-posed. This may be not true, however, when the functional depends on the final state, i.e. the problem is of Mayer's type

$$J(u(\cdot),x(\cdot),y(\cdot)) = g(x(T),y(T)).$$

In this case, using Lemma 3 we obtain:

Theorem 3.

Let the conditions A1 and A2 hold and the function $g(\cdot)$ be continuous. Define the problem

$$(P) \quad \hat{g} = \min g(x,y), \ (x,y) \in G(T),$$

where $G(T)$ is given in (8). Then

$$\hat{J}_\varepsilon \longrightarrow \hat{g} \text{ as } \varepsilon \longrightarrow 0$$

and

$$\hat{g} < \hat{J}_o.$$

The proper limit problem in this case is the problem (P) and the order reduction may be not well-posed.

5. Case $A_4 = 0$.

Consider the system (2) with $A_4 = 0$

$$(9) \quad \begin{aligned} \dot{x} &= A_1x + A_2y + Bu, \ x(0) = x^o, \\ \varepsilon\dot{y} &= A_3x + \varepsilon A_4'y, \qquad y(0) = y^o. \end{aligned}$$

In this case the reduced model

$$(10i) \quad \dot{x} = A_1x + A_2y + Bu, \ x(0) = x^o,$$

$$(10ii) \quad 0 = A_3x$$

is not well-defined since one can not eliminate y from (10i). Under some conditions, however, the differential equation (10i) can be dropped and the proper limit model is simply (10ii).

Denoting

$$\sqrt{\varepsilon} = \lambda, \ z = \lambda y, \ x = x_1 + x_2$$

one can rewrite (9) in the following equivalent form

$$\begin{aligned}
\dot{x}_1 &= A_1x_1 && + Bu, && x_1(0) = x^o, \\
\lambda\dot{x}_2 &= && \lambda A_2x_2 + A_2z, && x_2(0) = 0, \\
\lambda\dot{z} &= A_3x_1 + A_3x_2 && + \lambda A_4'z, && z(0) = \lambda y^o.
\end{aligned}$$

Suppose that the matrix

$$A = \begin{bmatrix} 0 & A_2 \\ A_3 & 0 \end{bmatrix}$$

satisfies the conditions in (A1) for the matrix A_4. The one can apply the results from the previous sections in order to justify the reduction from (9) to (10ii). Lemma 2 (with Remark 1) yields that if $u_\lambda(\cdot) \longrightarrow u_o(\cdot)$ as $\lambda \longrightarrow 0$ in the weak L_2 topology, then the corresponding solution $x_\lambda(\cdot)$ of (9) satisfies

$$x_\lambda(\cdot) \longrightarrow 0 \text{ as } \lambda \longrightarrow 0$$

L_2-weakly or even uniformly if $x^o = 0$ (the matrix A_3 is invertible). As an example consider the optimal control problem

$$\text{Minimize } J(u(\cdot),x(\cdot)) = \int_0^T (\phi(u(t)) + c(t)x(t))dt$$

subject to $u(\cdot) \in \mathcal{U}$ and (9),

where $\phi(\cdot)$ is continuous function. Then the infimal value $\hat{J}_\varepsilon$ satisfies

$$\hat{J}_\varepsilon \longrightarrow \hat{J} \text{ as } \varepsilon \longrightarrow 0,$$

where $\hat{J}$ is the optimal value of the following mathematical programming problem

$$\hat{J} = \min T\,\phi(u),\ u \in V$$

(compare with Theorem 2). If $x^o = 0$ a similar result can be obtained for more general nonlinear functionals.

For Mayer's problem with a functional

$$J(u(\cdot),x(\cdot)) = g(x(T))$$

assuming that exp(At) is periodic and $x^o = 0$, by Lemma 3 we conclude that

$$\hat{J}_\varepsilon \longrightarrow g(0) \text{ as } \varepsilon \longrightarrow 0.$$

If $x^o \neq 0$, $\hat{J}_\varepsilon$ may not possess a limit when $\varepsilon \longrightarrow 0$.

6. Two open problems.

Besides the natural generalizations there are two open questions the solution of which, in the authors' opinion, will be important for the further investigations of singularly perturbed problems in critical cases:

(i) To analyse the behaviour of the entire set of trajectories when $B_2 \neq 0$ (the limit reachable set is not bounded).

(ii) To define a limit problem for objective functionals containing both terminal and integral parts.

References

[1] Kokotovic P., Applications of singular perturbation techniques to control problems. SIAM Review, 1984, to appear.

[2] Dontchev A.L., Veliov V.M., Singular perturbation in Mayer's problem for linear systems. SIAM J. Contr. Optim. 21 (1983), pp. 566-581.

[3] ----------------, Singular perturbations in linear control systems with weakly coupled stable and unstable fast subsystems. J. Math. Anal. Appl., 1984, to appear.

[4] ----------------, Singular perturbation in linear differential inclusions - critical case. In: Parametric Optimization and Approximation, Eds. B. Brosowski and F. Deutsch, Birkhäuser 1984, to appear.

SENSITIVITY ANALYSIS IN NONLINEAR PROGRAMMING[1] UNDER SECOND ORDER ASSUMPTIONS

Anthony V. Fiacco
Department of Operations Research
The George Washington University
Washington, D.C. 20052

and

Jerzy Kyparisis
Department of Decision Sciences
Florida International University
Miami, Florida 33199

Abstract

In this paper basic results on sensitivity analysis in differentiable nonlinear programming are surveyed. Also, a simpler standard proof of a recent result due to Kojima is given.

1. Introduction

Since the classical sensitivity analysis results of Fiacco and McCormick [10], many new developments appeared in the literature in this area. General and now standard results under second order assumptions were subsequently obtained by Fiacco [5] and Robinson [22]. Additional results were published by Bigelow and Shapiro [3], Levitin [18], Armacost and Fiacco [1], and Fiacco [6]. More recently, extensions of these and many new results were obtained by Jittorntrum [12, 13], Kojima [16], Robinson [23,24,25], Spingarn [27,28], and Edahl [4].

In this paper we provide a survey of many of these results. We start in Section 3 with the results due to Fiacco [5], which hold under the strongest assumptions. In the following sections, we state other results under progressively weaker assumptions. In Section 5, we prove an important result due to Kojima [16], using only the standard tools of advanced calculus and employing some recent results of Robinson [25]. This approach was previously utilized by Edahl [4], and contrasts with Kojima's approach, which relies on advanced concepts of nonlinear analysis.

---

[1]This research was supported in part by the National Science Foundation.

Section 2 contains a review of second order optimality conditions and constraint qualifications used in the sequel.

2. Review of Second Order Optimality Conditions and Constraint Qualifications

A nonlinear programming (NLP) problem is of the form

$$\begin{aligned} &\underset{x \in E^n}{\text{minimize}} && f(x) \\ &\text{subject to} && g_i(x) \geq 0 \ , \ i = 1,\ldots,m \qquad \text{(P)} \\ & && h_j(x) = 0 \ , \ j = 1,\ldots,p \ , \text{ where} \\ & && f, g_i, h_j : E^n \to E^1 . \end{aligned}$$

It is assumed in this section that f, $\{g_i\}$ and $\{h_j\}$ are twice continuously differentiable around x^* . The Lagrangian associated with (P) is defined as:

$$L(x,u,w) = f(x) - \sum_{i=1}^{m} u_i g_i(x) + \sum_{j=1}^{p} w_j h_j(x) ,$$

$$\text{where} \quad u = (u_1,\ldots,u_m)^T \quad \text{and} \quad w = (w_1,\ldots,w_p)^T$$

are the Lagrange multiplier vectors associated with inequality and equality constraints $\{g_i\}$ and $\{h_j\}$, respectively. A point x_0 is a local minimum of (P) if x_0 is feasible, i.e.,

$$g_i(x_0) \geq 0 \ , \ i = 1,\ldots,m \ , \ h_j(x_0) = 0, \ j = 1,\ldots,p \ ,$$

and if there exists a neighborhood $N(x_0)$ of x_0 such that $f(x) \geq f(x_0)$ for all $x \in N(x_0)$ and feasible.

A general statement of first order necessary conditions for (local) optimality is given below.

Theorem 2.1 (Karush [14], Kuhn and Tucker [17]).

Suppose that x^* is a local minimum of (P) and that an appropriate constraint qualification (to be stipulated) holds at x^*. Then, the Karush-Kuhn-Tucker (KKT) conditions hold at x^* for (P), i.e., there exist Lagrange multiplier vectors u^* and w^* such that:

$$
\begin{aligned}
\nabla_x L(x^*,u^*,w^*) &= 0 , \\
u_i^* g_i(x^*) &= 0 , \quad i = 1,\dots,m , \\
g_i(x^*) &\geqslant 0 , \quad i = 1,\dots,m , \qquad \text{(KKT)} \\
h_j(x^*) &= 0 , \quad j = 1,\dots,p , \\
u_i^* &\geqslant 0 , \quad i = 1,\dots,m .
\end{aligned}
$$

There are many constraint qualifications which suffice for Theorem 2.1 to hold (see Fiacco and McCormick [10]). Two that are widely used later are:

(a) The Mangasarian-Fromovitz Constraint Qualification (MFCQ) holds at x^* if:

(i) the vectors $\{\nabla_x h_j(x^*),\ j = 1,\dots,p\}$ are linearly independent,

(ii) there is z such that

$$
\begin{aligned}
\nabla_x g_i(x^*)z &> 0 , \ i \in I(x^*) = \{i \mid g_i(x^*) = 0\} , \qquad \text{(MFCQ)} \\
\nabla_x h_j(x^*)z &= 0 , \ j = 1,\dots,p ; \ \text{and}
\end{aligned}
$$

(b) The Linear Independence condition (LI) holds at x^* if the vectors

$$\{\nabla_x g_i(x^*) ,\ i \in I(x^*) ,\ \nabla_x h_j(x^*) ,\ j = 1,\dots,p\} \qquad \text{(LI)}$$

are linearly independent.

It follows that (LI) implies (MFCQ), and also that (LI) implies the uniqueness of u^* and w^* in (KKT).

The next theorem states second order necessary conditions for (local) optimality.

<u>Theorem 2.2</u> (Fiacco and McCormick [10], McCormick [19]).

Suppose that x^* is a local minimum of (P) and that the Linear Independence condition holds at x^*. Then, the KKT conditions hold at x^* with associated unique Lagrange multiplier vectors u^* and w^*, and the additional Second Order Necessary Condition (SONC) holds at x^* with (u^*,w^*),

$$
\begin{aligned}
z^T\nabla_x^2 L(x^*,u^*,w^*)z &\geqslant 0 , \ \text{for all } z \ \text{ s.t.} \\
\nabla_x g_i(x^*)z &\geqslant 0 , \ \text{for all } i \in I(x^*) , \qquad \text{(SONC)} \\
\nabla_x g_i(x^*)z &= 0 , \ \text{for all } i \ \text{ s.t. } u_i^* > 0 , \\
\nabla_x h_j(x^*)z &= 0 , \ j = 1,\dots,p .
\end{aligned}
$$

By strengthening (SONC) one obtains the following standard second order sufficient conditions for (local) "strict" optimality (to be defined).

Theorem 2.3 (Pennisi [21], Fiacco and McCormick [10]).

Suppose that the KKT conditions hold at x^* for (P) with some Lagrange multiplier vectors u^* and w^*, and that the additional Second Order Sufficient Condition (SOSC) holds at x^* with (u^*,w^*),

$$\begin{aligned} z^T\nabla_x^2 L(x^*,u^*,w^*)z &> 0, \text{ for all } z \neq 0 \text{ s.t.} \\ \nabla_x g_i(x^*)z &\geq 0, \text{ for all } i \in I(x^*), \\ \nabla_x g_i(x^*)z &= 0, \text{ for all } i \text{ s.t. } u_i^* > 0, \\ \nabla_x h_j(x^*)z &= 0, \ j = 1,\dots,p. \end{aligned} \qquad \text{(SOSC)}$$

Then x^* is a strict local minimum of (P), i.e., $f(x) > f(x^*)$ for all feasible x in some neighborhood of x^*, where $x \neq x^*$.

Recently, Robinson [25] pointed out (also see Fiacco [8]) that x^* *need not* be an isolated (i.e., locally unique) local minimum under the KKT and SOSC assumptions, as indicated in Fiacco and McCormick [10, Thm. 4, p. 30]. He provides the following example in E^1.

Example 2.1 (Robinson [25]).

$$\underset{x}{\text{Minimize}} \qquad f(x) = \tfrac{1}{2}x^2$$

$$\text{subject to } h_1(x) = x^6\sin(1/x) = 0, \text{ where } h_1(0) := 0 .$$

One can easily verify that the assumptions of Theorem 2.3 are satisfied at $x^* = 0$. However, every point in the set $\{(n\pi)^{-1} \mid n = \pm 1, \pm 2, \dots\}$ is an isolated feasible point, and therefore also a local minimum. Thus, $x^* = 0$ is not an isolated local minimum.

Conditions sufficient for x^* to be an isolated local minimum are obtained by Robinson [25], by strengthening the assumptions of Theorem 2.3 in two ways.

Theorem 2.4 (Robinson [25]).

Suppose that the Karush-Kuhn-Tucker conditions hold at x^* for (P) with some u^* and w^* and that the Mangasarian-Fromovitz Constraint Qualification holds at x^*. Moreover, assume that the following General Second Order Sufficient Condition (GSOSC) holds at x^*,

(SOSC) holds at x^* with (u,w) for every (u,w) such that (x^*,u,w) satisfies the KKT conditions. (GSOSC)

Then x^* is an isolated local minimum of (P), i.e., there exists a neighborhood $N(x^*)$ of x^* such that x^* is the only local minimum of (P) in $N(x^*)$.

Note that if the LI condition is substituted for MFCQ in Theorem 2.4, then (GSOSC) coincides with (SOSC), since the Lagrange multiplier vectors u^* and w^* are unique. Thus, (SOSC) and (LI) imply that the hypotheses of Theorem 2.4 hold, since (LI) implies (MFCQ), as previously noted. Other conditions implying that a local minimum be isolated are under investigation by the authors.

3. Basic Sensitivity Results in Nonlinear Programming

A general parametric nonlinear programming problem is defined as:

$$\begin{aligned} \min_{x \in E^n} \quad & f(x,\varepsilon) \\ \text{subject to} \quad & g_i(x,\varepsilon) \geq 0, \quad i = 1,\dots,m, \\ & h_j(x,\varepsilon) = 0, \quad j = 1,\dots,p, \end{aligned} \qquad P(\varepsilon)$$

where $\varepsilon \in E^r$ is the perturbation parameter, $f, g_i, h_j: E^n \times E^r \to E^1$. It is assumed in this section that the functions f, $\{g_i\}$ and $\{h_j\}$ and their partial derivatives with respect to x are C^1 in (x,ε) in some neighborhood of (x^*,ε^*). The Lagrangian associated with $P(\varepsilon)$ is defined by:

$$L(x,u,w,\varepsilon) = f(x,\varepsilon) - u^T g(x,\varepsilon) + w^T h(x,\varepsilon) ,$$

where $g = (g_1,\dots,g_m)^T$, $h = (h_1,\dots,h_p)^T$. If $\varepsilon = \varepsilon^*$ is fixed, then all the definitions given in the previous section apply to problem $P(\varepsilon^*)$.

The following result was originally proved for a special class of parametric NLP problems by Fiacco and McCormick [10] and later for a general parametric NLP problem $P(\varepsilon)$ by Fiacco [5], and forms the basis for sensitivity analysis in nonlinear programming.

Theorem 3.1 (Fiacco [5]).

Suppose that the second order sufficient conditions for a local minimum of $P(\varepsilon^*)$ hold at x^* with associated Lagrange multiplier vectors u^* and x^* [i.e., (KKT) and (SOSC) hold at x^* with (u^*,w^*)], that the LI condition holds at x^* for $P(\varepsilon^*)$, and that the Strict Complementary Slackness condition (SCS) holds at x^* with respect to u^* for $P(\varepsilon^*)$, i.e.,

$$u_i^* > 0 \text{ when } g_i(x^*,\varepsilon^*) = 0 , \quad i = 1,\ldots,m .$$

Then,

(a) x^* is an isolated local minimum of $P(\varepsilon^*)$ and the associated Lagrange multiplier vectors u^* and w^* are unique;

(b) for ε in a neighborhood of ε^*, there exists a unique once continuously differentiable vector function $y(\varepsilon) = [x(\varepsilon),u(\varepsilon),w(\varepsilon)]^T$ satisfying the second order sufficient conditions for a local minimum of $P(\varepsilon)$ such that $y(\varepsilon^*) = (x^*,u^*,w^*)^T$ and, hence, $x(\varepsilon)$ is a locally unique local minimum of $P(\varepsilon)$ with associated unique Lagrange multiplier vectors $u(\varepsilon)$ and $w(\varepsilon)$;

(c) The Linear Independence and Strict Complementary Slackness conditions hold at $x(\varepsilon)$ for ε near ε^* .

Fiacco [5] also shows that the derivative of $y(\varepsilon)$ can be calculated by noting that the following system of Karush-Kuhn-Tucker equations will hold at $y(\varepsilon)$ for ε near ε^* under the assumptions of Theorem 3.1,

$$\begin{aligned} \nabla_x L[x(\varepsilon),u(\varepsilon),w(\varepsilon),\varepsilon] &= 0 \\ u_i(\varepsilon)g_i[x(\varepsilon),\varepsilon] &= 0 , \quad i = 1,\ldots,m , \\ h_j[x(\varepsilon),\varepsilon] &= 0 , \quad j = 1,\ldots,p . \end{aligned} \tag{3.1}$$

Since these assumptions imply that the Jacobian, $M(\varepsilon)$, of the system (3.1) with respect to (x,u,w) is nonsingular, one obtains

$$M(\varepsilon)\nabla_\varepsilon y(\varepsilon) = -N(\varepsilon) \tag{3.2}$$

and

$$\nabla_\varepsilon y(\varepsilon) = -M(\varepsilon)^{-1}N(\varepsilon) \tag{3.3}$$

where $N(\varepsilon)$ is the Jacobian of the system (3.1) with respect to ε .

The system (3.2) at $\varepsilon = \varepsilon^*$ can be written in the form

$$M^* \begin{bmatrix} \nabla_\varepsilon x(\varepsilon^*) \\ \nabla_\varepsilon u(\varepsilon^*) \\ \nabla_\varepsilon w(\varepsilon^*) \end{bmatrix} = -N^* \tag{3.4}$$

where

$$M^* = \begin{bmatrix} \nabla_x^2 L & -\nabla_x g_1^T, & \ldots, & -\nabla_x g_m^T & \nabla_x h_1^T, & \ldots, & \nabla_x h_p^T \\ u_1^* \nabla_x g_1 & g_1 & & 0 & & & \\ & & & & & 0 & \\ \vdots & & \ddots & & & & \\ u_m^* \nabla_x g_m & 0 & & g_m & & & \\ \nabla h_1 & & & & & & \\ \vdots & & 0 & & & 0 & \\ \nabla h_p & & & & & & \end{bmatrix} \tag{3.5}$$

and

$$N^* = \left[\nabla_{\varepsilon x}^2 L^T, \ u_1^* \nabla_\varepsilon g_1^T, \ldots, u_m^* \nabla_\varepsilon g_m^T, \ \nabla_\varepsilon h_1^T, \ldots, \nabla_\varepsilon h_p^T \right]^T \tag{3.6}$$

are evaluated at $(x^*, u^*, w^*, \varepsilon^*)$.

The next theorem, due to McCormick [19], shows that the conditions imposed in Theorem 3.1 are also essentially necessary (under appropriate regularity assumptions) for the invertibility of the Jacobian matrix M^*.

Theorem 3.2 (McCormick [19]).

Suppose that the second order necessary conditions for a local minimum of $P(\varepsilon^*)$ hold at x^* with associated Lagrange multiplier vectors u^* and w^* [i.e., (KKT) and (SONC) hold at x^* with (u^*, w^*)].

Then, the Jacobian matrix M^* given by (3.5) is invertible if and only if the SOSC, LI and SCS conditions hold at x^* with (u^*, w^*) for $P(\varepsilon^*)$.

4. Sensitivity Analysis Without the Strict Complementarity Slackness Assumption

This section presents the results of Jittorntrum [12,13] and Robinson [24] which extend the results of the previous section by relaxing the SCS assumption and strengthening the standard second order sufficient conditions for (local) strict optimality. The differentiability assumptions on f, $\{g_i\}$ and $\{h_j\}$ are the same as in Section 3.

Theorem 4.1 (Jittorntrum [12,13], Robinson [24]).

Suppose that the Karush-Kuhn-Tucker conditions hold at x^* for $P(\varepsilon^*)$ with some Lagrange multiplier vectors u^* and w^*, that the additional Strong Second Order Sufficient Condition (SSOSC) holds at x^* with (u^*,w^*) ,

$$\begin{aligned} z^T\nabla_x^2 L(x^*,u^*,w^*,\varepsilon^*)z > 0 \text{ for all } z \neq 0 \text{ s.t.} \\ \nabla_x g_i(x^*,\varepsilon^*)z = 0 \text{ for all } i \text{ s.t. } u_i^* > 0 , \\ \nabla_x h_j(x^*,\varepsilon^*)z = 0 \ j = 1,\ldots,p \end{aligned} \qquad \text{(SSOSC)}$$

and that the LI condition holds at x^* for $P(\varepsilon^*)$.

Then,

(a) x^* is an isolated local minimum of $P(\varepsilon^*)$ and the associated Lagrange multiplier vectors u^* and w^* are unique;

(b) for ε in a neighborhood of ε^*, there exists a unique continuous vector function $y(\varepsilon) = [x(\varepsilon),u(\varepsilon),w(\varepsilon)]^T$ satisfying the Strong Second Order Sufficient Conditions (i.e., (KKT) and (SSOSC)) for a local minimum of $P(\varepsilon)$ such that $y(\varepsilon^*) = (x^*,u^*,w^*)^T$ and, hence, $x(\varepsilon)$ is a locally unique local minimum of $P(\varepsilon)$ with associated unique Lagrange multiplier vectors $u(\varepsilon)$ and $w(\varepsilon)$;

(c) the Linear Independence condition holds at $x(\varepsilon)$ for ε near ε^* ;

(d) there exist $t > 0$ and $d > 0$ such that for all ε with $\|\varepsilon - \varepsilon^*\| < d$, it follows that

$$\|y(\varepsilon) - y(\varepsilon^*)\| \leqslant t\|\varepsilon - \varepsilon^*\| .$$

The following system of inequalities and equations in $(\mathring{x},\mathring{u},\mathring{w})$ is a generalization of the system (3.4) and was first considered by Bigelow and Shapiro [3]:

$$\nabla_x^2 L(x^*,u^*,w^*,\varepsilon^*)\mathring{x} - \sum_{i=1}^{m} \mathring{u}_i \nabla_x g_i(x^*,\varepsilon^*)^T$$

$$+ \sum_{j=1}^{p} \mathring{w}_j \nabla_x h_j(x^*,\varepsilon^*)^T = -\nabla_{\varepsilon x}^2 L(x^*,u^*,w^*,\varepsilon^*)v$$

$$\nabla_x g_i(x^*,\varepsilon^*)\mathring{x} = -\nabla_\varepsilon g_i(x^*,\varepsilon^*)v, \qquad i \in I_1^* \qquad (4.1)$$

$$\nabla_x g_i(x^*,\varepsilon^*)\mathring{x} \geq -\nabla_\varepsilon g_i(x^*,\varepsilon^*)v\ , \qquad i \in I_2^*$$

$$\mathring{u}_i[\nabla_x g_i(x^*,\varepsilon^*)\mathring{x} + \nabla_\varepsilon g_i(x^*,\varepsilon^*)v] = 0\ , \qquad i \in I_2^*$$

$$\mathring{u}_i = 0\ , \quad i \notin I^*\ ; \qquad \mathring{u}_i \geq 0\ , \qquad i \in I_2^*$$

$$\nabla_x h_j(x^*,\varepsilon^*)\mathring{x} = -\nabla_\varepsilon h_j(x^*,\varepsilon^*)v\ , \qquad j = 1,\ldots,p\ .$$

where

$$I^* = \{i = 1,\ldots,m \mid g_i(x^*,\varepsilon^*) = 0\}$$

$$I_1^* = \{i \in I^* \mid u_i^* > 0\}\ ,\ I_2^* = \{i \in I^* \mid u_i^* = 0\}\ .$$

The next result was recently obtained by Jittorntrum [13]. An attempt in this direction was also made earlier by Bigelow and Shapiro [3].

Theorem 4.2 (Jittorntrum [13]).

Suppose that the assumptions of Theorem 4.1 are satisfied and let $v \neq 0$. Then the system of inequalities and equations (4.1) for the family of perturbed problems $P(\varepsilon^*+av)$, $a \geq 0$, has a unique solution $(\mathring{x},\mathring{u},\mathring{w})$. Furthermore, $\mathring{x}$, $\mathring{u}$ and $\mathring{w}$ are the one-sided directional derivatives of $x(\varepsilon)$, $u(\varepsilon)$ and $w(\varepsilon)$ at ε^* in the direction v, respectively.

5. Sensitivity Analysis Without the Linear Independence Assumption

The results of Section 4 are further extended in this section by substituting the weaker MFCQ condition for the LI condition and strengthening the strong second order sufficient conditions for (local) strict optimality. Kojima [16] obtained the main result of this section with some additional characterizations by making extensive use of the degree theory of continuous maps (see, e.g., Ortega and Rheinboldt [20]) and basic results for piecewise continuously differentiable (PC^1) maps (see Kojima [15]).

Our goal is to obtain this result for the first time by utilizing the more standard tools of advanced calculus. This approach was previously adopted by Edahl [4]. Other related results were obtained under similar assumptions by Levitin [18]. Before proving the main theorem we state two useful results due to Robinson [25] which are of independent interest.

It is assumed in this and the next section that f, $\{g_i\}$ and $\{h_j\}$ are C^2 in x for every ε and that $\nabla_x f$, $\{\nabla_x g_i\}$ and $\{\nabla_x h_j\}$ are continuous in (x,ε) near (x^*,ε^*).

<u>Theorem 5.1</u> (Robinson [25]).

Suppose that the second order sufficient conditions for a local minimum of $P(\varepsilon^*)$ hold at x^* with associated Lagrange multiplier vectors u^* and w^* [i.e., (KKT) and (SOSC) hold at x^* with (u^*,w^*)], and that the MFCQ holds at x^* for $P(\varepsilon^*)$.

Then, for each neighborhood $N(x^*)$ of x^* there is a neighborhood $N(\varepsilon^*)$ of ε^* such that there exists a local minimum of $P(\varepsilon)$ in $N(x^*)$ for each ε in $N(\varepsilon^*)$.

The following notions can be found in Berge [2]. Let $\Gamma: D \to R$, $D \subset E^r$, $R \subset E^s$ be a point-to-set map associating with every point $z \in D$ a (possibly empty) subset $\Gamma(z) \subset R$. The map Γ is called upper semicontinuous at $z_0 \in D$ if for any open set $W \subset \Gamma(z_0)$ there is an open set V with $z_0 \in V$ such that for all $z \in V$, $\Gamma(z) \subset W$. It is called upper semicontinuous if for all $z \in D$ it is upper semicontinuous at z and $\Gamma(z)$ is <u>compact</u>.

Note that if Γ is upper semicontinuous and single-valued, then it is a continuous function.

The next result generalizes earlier results of Levitin [18] and Gauvin [11].

<u>Theorem 5.2</u> (Robinson [25]).

If the MFCQ holds at x^* for $P(\varepsilon^*)$, then there exist neighborhoods N_1 of x^* and N_2 of ε^*, such that if $K: N_1 \times N_2 \to E^m \times E^p$ and $SP: N_2 \to N_1$ are point-to-set maps of Lagrange multiplier vectors and stationary points, respectively, defined by

$$K(x,\varepsilon) = \{(u,w) \in E^m \times E^p \mid \nabla_x L(x,u,w,\varepsilon) = 0 ,$$

$$u_i g_i(x,\varepsilon) = 0 , u_i \geqslant 0 , i = 1,\ldots,m\}$$

for $(x,\varepsilon) \in N_1 \times N_2$,

$$SP(\varepsilon) = \{x \in N_1 \mid \text{ for some } (u,w) \in K(x,\varepsilon)\}$$

for $\varepsilon \in N_2$,

then K and SP are upper semicontinuous and there is a compact set $S \subset E^m \times E^p$ such that $K(x,\varepsilon) \subset S$ for all $(x,\varepsilon) \in N_1 \times N_2$.

We finally state our main result.

Theorem 5.3 (Kojima [16]).

Suppose that the Karush-Kuhn-Tucker conditions hold at x^* for $P(\varepsilon^*)$ with some Lagrange multiplier vectors $\bar{u}^*$ and $\bar{w}^*$, that the additional General Strong Second Order Sufficient Condition (GSSOSC) holds at x^* , defined as

$$\text{(SSOSC) holds at } x^* \text{ with } (u,w) \quad \text{for every } (u,w) \in K(x^*,\varepsilon^*) , \tag{GSSOSC}$$

and that the MFCQ holds at x^* for $P(\varepsilon^*)$. Then,

(a) x^* is an isolated local minimum of $P(\varepsilon^*)$ and the set $K(x^*,\varepsilon^*)$ is compact and convex;

(b) there are neighborhoods $N(x^*)$ of x^* and $N(\varepsilon^*)$ of ε^* such that for ε in $N(\varepsilon^*)$ there exists a unique continuous vector function $x(\varepsilon)$ in $N(x^*)$ satisfying the KKT conditions with some $[u(\varepsilon),w(\varepsilon)] \in K(x(\varepsilon),\varepsilon)$ and the General Strong Second Order Sufficient Condition such that $x(\varepsilon^*) = x^*$, and hence $x(\varepsilon)$ is the locally unique local minimum of $P(\varepsilon)$ in $N(x^*)$;

(c) the MFCQ holds at $x(\varepsilon)$ for ε in $N(\varepsilon^*)$.

Proof:

Part (a). The first part follows immediately from Theorem 2.4. Since $K(x,\varepsilon)$ is closed, Theorem 5.2 implies that $K(x^*,\varepsilon^*)$ is compact. The convexity of $K(x^*,\varepsilon^*)$ follows trivially.

Part (b). We first prove the uniqueness of stationary points $x(\varepsilon)$ in some $N(x^*)$ (assuming that they exist), by contradiction.

Let $\{\eta_k\}$ and $\{d_k\}$ be two arbitrary sequences of positive numbers such that $\eta_k \downarrow 0$, $d_k \downarrow 0$. Denote by $C(z,r)$ an open ball around $z \in E^s$ with radius $r > 0$. Suppose that there exist sequences $\{\varepsilon_k\}$, $\{x_k\}$ and $\{z_k\}$ such that $\varepsilon_k \varepsilon C(\varepsilon^*, d_k)$ and $x_k, z_k \ \varepsilon \ SP(\varepsilon_k)$, where $x_k \neq z_k$, $x_k, z_k \ \varepsilon \ C(x^*, \eta_k)$. Without loss of generality, assume that all the constraints g_i are binding at (x^*, ε^*).

Let $N_0 = \{0,1,2,\ldots\}$, $J = \{k \ \varepsilon \ N_0 \mid f(x_k, \varepsilon_k) \geqslant f(z_k, \varepsilon_k)\}$, $\bar{J} = N_0 - J$. Since at least one of the sets $J, \bar{J}$ is infinite, we can assume, for example, that J is infinite, relabel the sequences, and have $f(x_k, \varepsilon_k) \geqslant f(z_k, \varepsilon_k)$ for all k. Let $I_k = I(x_k, \varepsilon_k) = \{i = 1,\ldots,m \mid g_i(x_k, \varepsilon_k) = 0\}$. Since $I_k \subset \{1,\ldots,m\}$ for all k and the number of subsets of $\{1,\ldots,m\}$ is finite, we can find an infinite set K and a set $\bar{I} \subset \{1,\ldots,m\}$ such that $I_k = \bar{I}$ for all $k \ \varepsilon \ K$. Again, by relabeling the sequences we have $I_k = \bar{I}$ for all k.

Define $a_k = \| z_k - x_k \|$, $s_k = (z_k - x_k)/a_k$, so that $z_k = x_k + a_k s_k$ with $a_k > 0$, $a_k \to 0$ (since $x_k, z_k \to x^*$). Since $\| s_k \| = 1$, all k, there exists a convergent subsequence of s_k, and after relabeling we can write for some s, $s_k \to s$, $\| s \| = 1$. We have that for all k

$$
\begin{aligned}
&f(z_k, \varepsilon_k) \leqslant f(x_k, \varepsilon_k) \ , \\
&g_i(z_k, \varepsilon_k) \geqslant g_i(x_k, \varepsilon_k) = 0 \ , \ i \ \varepsilon \ \bar{I} \ , \\
&h_j(z_k, \varepsilon_k) = h_j(x_k, \varepsilon_k) = 0 \ , \ j = 1,\ldots,p \ .
\end{aligned} \tag{5.1}
$$

By Taylor's theorem there exist $0 \leqslant v_i^k \leqslant 1$, for all k such that

$$
\begin{aligned}
&f(z_k, \varepsilon_k) = f(x_k, \varepsilon_k) + \nabla_x f(x_k + v_0^k a_k s_k, \varepsilon_k) a_k s_k \ , \\
&g_i(z_k, \varepsilon_k) = g_i(x_k, \varepsilon_k) + \nabla_x g_i(x_k + v_i^k a_k s_k, \varepsilon_k) a_k s_k \ , \ i \ \varepsilon \ \bar{I} \\
&h_j(z_k, \varepsilon_k) = h_j(x_k, \varepsilon_k) + \nabla_x h_j(x_k + v_{m+j}^k a_k s_k, \varepsilon_k) a_k s_k \ , \ j=1,\ldots,p.
\end{aligned} \tag{5.2}
$$

Combining (5.1) and (5.2) and the fact that $a_k > 0$, one obtains for all k that

$$
\begin{aligned}
&\nabla_x f(x_k + v_0^k a_k s_k, \varepsilon_k) s_k \leqslant 0 \ , \\
&\nabla_x g_i(x_k + v_i^k a_k s_k, \varepsilon_k) s_k \geqslant 0 \ , \ i \ \varepsilon \ \bar{I} \\
&\nabla_x h_j(x_k + v_{m+j}^k a_k s_k, \varepsilon_k) s_k = 0 \ , \ j = 1,\ldots,p \ .
\end{aligned} \tag{5.3}
$$

Taking the limits as $k \to \infty$ in (5.3) yields

$$\begin{aligned} \nabla_x f(x^*,\varepsilon^*)s &\leqslant 0 , \\ \nabla_x g_i(x^*,\varepsilon^*)s &\geqslant 0 , \quad i \in \bar{I} , \\ \nabla_x h_j(x^*,\varepsilon^*)s &= 0 , \quad j = 1,\ldots,p . \end{aligned} \tag{5.4}$$

Define u^k and w^k to be the Lagrange multiplier vectors at x_k for $P(\varepsilon_k)$ (we can choose arbitrary ones). By Theorem 5.2 the point-to-set map K is (locally) upper semicontinuous and uniformly compact, i.e., $K(x,\varepsilon) \subset S$ for all $(x,\varepsilon) \in N_1 \times N_2$ and some compact set $S \subset E^m \times E^p$. Thus, we can assume that $(u^k,w^k) \to (u^*,w^*)$ and by upper semicontinuity $(u^*,w^*) \in K(x^*,\varepsilon^*)$, i.e., (x^*,u^*,w^*) satisfies the KKT conditions for $P(\varepsilon^*)$.

Let $I_0 = \{i = 1,\ldots,m \mid u_i^* > 0\}$. It is easy to see that $I_0 \subset \bar{I}$ (otherwise for some $i_0 \in I_0 - \bar{I}$ and large k, $u_{i_0}^k g_{i_0}(x_k,\varepsilon_k) > 0$, a contradiction to complementarity). Since (x^*,u^*,w^*) satisfies the KKT conditions for $P(\varepsilon^*)$,

$$\nabla_x f(x^*,\varepsilon^*) = \sum_{i\varepsilon I_0} u_i^* \nabla_x g_i(x^*,\varepsilon^*) - \sum_{j=1}^{p} w_j^* \nabla_x h_j(x^*,\varepsilon^*) .$$

This implies that

$$\nabla_x f(x^*,\varepsilon^*)s = \sum_{i\varepsilon I_0} u_i^* \nabla_x g_i(x^*,\varepsilon^*)s - \sum_{j=1}^{p} w_j^* \nabla_x h_j(x^*,\varepsilon^*)s .$$

However, since $u_i^* > 0$, $i \in I_0$, $I_0 \subset \bar{I}$, in view of (5.4)

$$\begin{aligned} \nabla_x f(x^*,\varepsilon^*)s &= 0 , \\ \nabla_x g_i(x^*,\varepsilon^*)s &= 0 , \quad i \in I_0 \\ \nabla_x h_j(x^*,\varepsilon^*)s &= 0 , \quad j = 1,\ldots,p. \end{aligned} \tag{5.5}$$

Using Taylor's theorem once more, one can write for some $0 \leqslant t_i^k \leqslant 1$ and all k

$$f(z_k,\varepsilon_k) = f(x_k,\varepsilon_k) + \nabla_k f(x_k,\varepsilon_k)a_k s_k + \tfrac{1}{2}(a_k)^2 s_k^T \nabla_x^2 f(x_k+t_0^k a_k s_k,\varepsilon_k)s_k ,$$

$$g_i(z_k,\varepsilon_k) = g_i(x_k,\varepsilon_k) + \nabla_x g_i(x_k,\varepsilon_k)a_k s_k$$
$$+ \tfrac{1}{2}(a_k)^2 s_k^T \nabla_x^2 g_i(x_k+t_i^k a_k s_k,\varepsilon_k)s_k \ , \ i \in \bar{I} \ ,$$

$$h_j(z_k,\varepsilon_k) = h_j(x_k,\varepsilon_k) + \nabla_x h_j(x_k,\varepsilon_k)a_k s_k$$
$$+ \tfrac{1}{2}(a_k)^2 s_k^T \nabla_x^2 h_j(x_k+t_{m+j}^k a_k s_k,\varepsilon_k)s_k \ , \ j = 1,\ldots,p \ . \tag{5.6}$$

Multiplying the corresponding equations in (5.6) by $-u_i^k$ and w_j^k, adding them, and using (5.1) and the KKT conditions at x_k for $P(\varepsilon_k)$, one obtains for all k

$$s_k^T \left[\nabla_x^2 f(x_k+t_0^k a_k s_k,\varepsilon_k) - \sum_{i\varepsilon\bar{I}} u_i^k \nabla_x^2 g_i(x_k+t_i^k a_k s_k,\varepsilon_k) + \sum_{j=1}^{p} w_j^k \nabla_x^2 h_j(x_k+t_{m+j}^k a_k s_k,\varepsilon_k) \right] s_k \leq 0 \tag{5.7}$$

Taking the limit in (5.7) as $k \to \infty$ and recalling that $I_0 \subset \bar{I}$ yields

$$s^T \left[\nabla_x^2 f(x^*,\varepsilon^*) - \sum_{i\varepsilon I_0} u_i^* \nabla_x^2 g_i(x^*,\varepsilon^*) + \sum_{j=1}^{p} w_j^* \nabla_x^2 h_j(x^*,\varepsilon^*) \right] s \leq 0$$

which together with (5.5) contradicts the satisfaction of the GSSOSC condition at x^*.

Thus, we have proven that there are neighborhoods $N(x^*)$ of x^* and $N_0(\varepsilon^*)$ of ε^* such that for ε in $N_0(\varepsilon^*)$ the stationary points of $P(\varepsilon)$ in $N(x^*)$ are unique if they exist. But, by Theorem 5.1, there exists a local minimum $x(\varepsilon)$ of $P(\varepsilon)$ in $N(x^*)$ for all ε in some $N(\varepsilon^*) \subset N_0(\varepsilon^*)$ and $x(\varepsilon)$ is also a stationary point of $P(\varepsilon)$ [since Part (c), the MFCQ holds at $x(\varepsilon)$ for $P(\varepsilon)$ and $\varepsilon \in N(x^*)$ has been shown [Robinson, 25]].

Therefore, we can conclude that there is a unique local minimum (and stationary point) $x(\varepsilon)$ of $P(\varepsilon)$ in $N(x^*)$ for all ε in $N(\varepsilon^*)$. Also, by Theorem 5.2, $SP(\varepsilon) = \{x(\varepsilon)\}$ is an upper semicontinuous point-to-set map in $N(\varepsilon^*)$ and thus $x(\varepsilon)$ is a continuous vector function in $N(\varepsilon^*)$. Finally, Kojima [16, Lemma 7.5] proved in a standard way that the GSSOSC condition is preserved under small perturbations, if the assumptions of the theorem are satisfied. Thus, part (b) is proved.

6. Additional Extensions of Sensitivity Analysis

The results of Section 4 were obtained under the KKT, SSOSC and LI conditions, which are weaker than the KKT, SOSC, SCS and LI conditions imposed in Section 3. These latter assumptions can also be relaxed to the following set of assumptions: KKT, GSOSC, GSCS and MFCQ, where the General Strict Complementarity Slackness condition (GSCS) is said to hold at x^* for $P(\varepsilon^*)$ if SCS holds at x^* w.r.t. (u,w) for every $(u,w) \in K(x^*,\varepsilon^*)$. This generalization turns out, however, to be spurious since, as it will be shown later, the last two sets of assumptions are equivalent. This observation was made previously by Edahl [4].

Proposition 6.1.

Suppose that x^* is feasible for $P(\varepsilon^*)$ and that the KKT, GSCS and MFCQ conditions hold at x^*. Then $K(x^*,\varepsilon^*) = \{(u^*,w^*)\}$ and the LI condition holds at x^*.

Proof:

Since the KKT and GSCS conditions hold at x^* for $P(\varepsilon^*)$, for any optimal Lagrange multiplier vectors u and w we have

$$\nabla_x f(x^*,\varepsilon^*) - u^T \nabla_x g(x^*,\varepsilon^*) + w^T \nabla_x h(x^*,\varepsilon^*) = 0 ,$$

$$u_i > 0 , \; i = 1,\ldots,m ,$$

where it is assumed for simplicity that all the constraints are binding. If (u^1,w^1), $(u^2,w^2) \in K(x^*,\varepsilon^*)$, then it is easy to see that $t_1(u^1,w^1) + t_2(u^2,w^2) \in K(x^*,\varepsilon^*)$ for all t_1,t_2 such that $t_1 + t_2 = 1$ and $t_1u_i^1 + t_2u_i^2 > 0$, $i = 1,\ldots,m$. If $u^1 \neq u^2$, then for some t_{10}, t_{20} we will have $u_i^0 = t_{10}u_i^1 + t_{20}u_i^2 \geqslant 0$, $i = 1,\ldots,m$ and $u_k^0 = t_{10}u_k^1 + t_{20}u_k^2 = 0$ for some $k \in \{1,2,\ldots,m\}$. But this contradicts the GSCS assumption; thus $u^1 = u^2$. Moreover, since the MFCQ holds, the KKT conditions imply the uniqueness of the Lagrange multiplier vector w; hence $K(x^*,\varepsilon^*) = \{(u^*,w^*)\}$ ..

Suppose now that the LI condition does not hold at x^*. Then there exists a vector $0 \neq a \in E^{m+p}$ such that $a^TP = 0$, where $P = [-\nabla_x g(x^*,\varepsilon^*)^T , \nabla_x h(x^*,\varepsilon^*)^T]^T$. This implies that

$$\nabla_x f(x^*,\varepsilon^*) + [((u^*)^T,(w^*)^T) + ta^T]P = 0$$

for all t, and thus for small t, $z(t) = (u^*,w^*) + ta$ satisfies the KKT and SCS conditions. This, however, contradicts the fact that

$K(x^*,\varepsilon^*)$ is a singleton. Q.E.D.

The next result easily follows from Proposition 6.1.

Proposition 6.2.

The following two sets of assumptions are equivalent:

(a) the KKT, SOSC, SCS and LI conditions hold at x^* for $P(\varepsilon^*)$;

(b) the KKT, GSOSC, GSCS and MFCQ conditions hold at x^* for $P(\varepsilon^*)$.

In Sections 3 through 5, sensitivity analysis was carried out under progressively weaker assumptions. However, it should be clear that, in the absence of inequality constraints, all of these sets of conditions reduce to the KKT, SOSC, and LI conditions. Additional results were recently obtained by Robinson [25] under even more general assumptions than those adopted in Section 5. In order to state his main result, we need to introduce the following definitions (see Berge [2]).

A point-to-set map $\Gamma: D \to R$, $D \subset E^r$, $R \subset E^s$ is called lower semicontinuous at $z_0 \in D$ if for any $v \in \Gamma(z_0)$ and for any open set W with $v \in W$ there exists an open set V with $z_0 \in V$ such that for all $z \in V$, $\Gamma(z) \cap W \neq \emptyset$. Upper semicontinuity of Γ at a point was defined previously. Finally, Γ is called continuous at $z_0 \in D$ if it is both upper and lower semicontinuous at z_0 .

Theorem 6.3 (Robinson [25]).

Suppose that the Karush-Kuhn-Tucker conditions hold at x^* for $P(\varepsilon^*)$ with some Lagrange multiplier vectors u^* and w^* , that the additional General Second Order Sufficient Condition holds at x^* for $P(\varepsilon^*)$ (see Section 2), and that the MFCQ holds at x^*. Then,

(a) x^* is an isolated local minimum of $P(\varepsilon^*)$ and the set $K(x^*,\varepsilon^*)$ is compact and convex;

(b) there exist neighborhoods $N(x^*)$ of x^* and $N(\varepsilon^*)$ of ε^* such that if the point-to-set map $LS: N(\varepsilon^*) \to N(x^*)$ is defined by
$$LS(\varepsilon) = \{x \in N(x^*) \mid x \text{ is a local minimum of } P(\varepsilon)\} ,$$
then the point-to-set map $SP \cap N(x^*)$ [defined by $(SP \cap N(x^*))(\varepsilon) = SP(\varepsilon) \cap N(x^*)$] is continuous at ε^* and for each $\varepsilon \in N(\varepsilon^*)$ has $\emptyset \neq LS(\varepsilon) \subset SP(\varepsilon)$;

(c) the MFCQ holds at all points in $SP(\varepsilon) \cap N(x^*)$ for ε in $N(\varepsilon^*)$.

The following example provided by Robinson [24] shows that under the assumptions of Theorem 6.3 the map LS may not be single-valued near $\varepsilon^* = 0$ and that the inclusion of LS in SP may be strict.

Example 6.1 (Robinson [24]).

$$\begin{aligned} \text{Minimize}_x \quad f(x,\varepsilon) &= (x_1-\varepsilon)^2 - x_2^2 \\ \text{subject to } g_1(x,\varepsilon) &= x_1 - 2x_2 \geqslant 0 \\ g_2(x,\varepsilon) &= x_1 + 2x_2 \geqslant 0 \quad . \end{aligned}$$

For $\varepsilon^* = 0$, $x^* = (0,0)^T$ is the unique stationary point of the problem with associated Lagrange multipliers $u_1^* = 0$, $u_2^* = 0$. The LI condition holds at x^* since both constraints are binding at x^* and

$$\nabla_x g_1(x^*,\varepsilon^*) = (1,-2) \ , \quad \nabla_x g_2(x^*,\varepsilon^*) = (1,2) \ .$$

Also, the SOSC condition holds at x^* , as can be easily checked.

Thus, the assumptions of Theorem 6.3 are satisfied at x^* . However, it turns out that for all $\varepsilon > 0$

$$LS(\varepsilon) = \{(4\varepsilon/3,\ 2\varepsilon/3)^T\ ,\ (4\varepsilon/3,\ -2\varepsilon/3)^T$$

and

$$SP(\varepsilon) = LS(\varepsilon) \cup \{(\varepsilon,0)^T\}$$

This example shows that even when the stronger conditions, SOSC and LI, hold at x^* for $P(\varepsilon^*)$, the local minima may not be unique near $\varepsilon = \varepsilon^*$.

7. Differentiability of the Optimal Value Function

In differential stability results (see the survey in Fiacco [8]) one employs the standard definition of the optimal value function f^*

$$f^*(\varepsilon) = \begin{cases} \inf_x \{f(x,\varepsilon) \mid x \in R(\varepsilon)\}, & \text{if } R(\varepsilon) \neq \emptyset \\ +\infty & , \text{if } R(\varepsilon) = \emptyset \end{cases} \tag{7.1}$$

where $R(\varepsilon) = \{x \mid g_i(x,\varepsilon) \geqslant 0,\ i = 1,\dots,m;\ h_j(x,\varepsilon) = 0,\ j = 1,\dots,p\}$, and usually assumes some kind of compactness of the sets $R(\varepsilon)$. This definition, however, is not suitable when dealing with results having a local character as those given in the previous sections. In this case, usually (see [6,7,8]) the "local" optimal value function f_ℓ^* is defined by

$$f_\ell^*(\varepsilon) = f[x(\varepsilon),\varepsilon] \ , \tag{7.2}$$

where $x(\varepsilon)$ is an isolated local minimum of $P(\varepsilon)$. It is clear that in general $f^* \leqslant f_\ell^*$ unless the problems $P(\varepsilon)$ are convex. Nevertheless, the gap between results involving f^* and f_ℓ^* can be locally bridged under certain assumptions, using the notion of a "restricted" optimal value function f_N^* defined for an appropriate choice of a neighborhood N as

$$f_N^*(\varepsilon) = \begin{cases} \inf_x\{f(x,\varepsilon) \mid x \in R(\varepsilon) \cap N\} , & \text{if } R(\varepsilon) \cap N \neq \emptyset \\ +\infty , & \text{if } R(\varepsilon) \cap N = \emptyset \end{cases} \tag{7.3}$$

i.e., f_N^* is the optimal value function of the restricted problem

$$\min_x f(x,\varepsilon) \qquad \text{s.t.} \quad x \in R(\varepsilon) \cap N \qquad\qquad P_N(\varepsilon)$$

Proposition 7.1.

Suppose that the assumptions of Theorem 5.3 are satisfied, i.e., that the GSSOSC and MFCQ conditions hold at x^* for $P(\varepsilon^*)$. Then, there exist a neighborhood N_0 of ε^* and a closed ball B around x^* such that for ε in N_0

$$f_B^*(\varepsilon) = f[x(\varepsilon),\varepsilon] ,$$

where $x(\varepsilon) \in \text{int}B$ (defined in Theorem 5.3) is the unique global minimum of $P_B(\varepsilon)$.

Proof:

Consider a closed ball $B = \{x \mid \|x-x^*\| \leqslant r\}$ with some $r > 0$ contained in $N(x^*)$ (defined in Theorem 5.3). Since x^* is a strict local minimum of $P(\varepsilon^*)$ and the set $T = \{x \mid \|x-x^*\| = r\}$ is compact, $\inf_x\{f(x,\varepsilon^*) \mid x \in T\} = c^* > f(x^*,\varepsilon^*)$. Denote $c(\varepsilon) = \inf_x\{f(x,\varepsilon) \mid x \in T\}$. Since f is continuous and T is compact, it follows that c is continuous (Berge [2]). Thus, $c(\varepsilon) \geqslant d > f(x^*,\varepsilon^*)$ for some d and all ε in some neighborhood N of ε^* , and consequently $f(x,\varepsilon) \geqslant d > f[x(\varepsilon),\varepsilon]$ for all $x \in T$ and ε in some neighborhood N_0 of ε^* in view of the continuity of $x(\varepsilon)$. Since, by Theorem 5.3, $x(\varepsilon)$ is the unique local minimum of $P(\varepsilon)$ in $N(x^*)$ and any problem $P_B(\varepsilon)$ possesses a global minimum by compactness of $R(\varepsilon) \cap B$, the above inequality implies that $x(\varepsilon)$ is the unique global minimum of $P_B(\varepsilon)$.

Proposition 7.1 shows that under the assumptions of Theorem 5.3 the local optimal value function f_ℓ^* is actually the standard optimal value function f^* restricted to a ball around x^*.

The following result was obtained in Armacost and Fiacco [1] and in Fiacco [6] as a consequence of Theorem 3.1.

Theorem 7.2 (Armacost and Fiacco [1], Fiacco [6]).

If the assumptions of Theorem 3.1 are satisfied, i.e., if the KKT SOSC, SCS and LI conditions hold at x^* for $P(\varepsilon)$ then, in a neighborhood of $\varepsilon = \varepsilon^*$, the local optimal value function f^*_ℓ is twice continuously differentiable and:

(a) $f^*_\ell(\varepsilon) = L[x(\varepsilon), u(\varepsilon), w(\varepsilon), \varepsilon]$; (7.4)

(b) $\nabla_\varepsilon f^*_\ell(\varepsilon) = \nabla_\varepsilon L(x,u,w,\varepsilon) \Big|_{[x(\varepsilon),u(\varepsilon),w(\varepsilon),\varepsilon]}$; (7.5)

$$\text{(c)}\quad \nabla^2_\varepsilon f^*_\ell(\varepsilon) = \nabla^2_{\varepsilon x} L(x,u,w,\varepsilon)^T \nabla_\varepsilon x(\varepsilon) - \nabla_\varepsilon g(x,\varepsilon)^T \nabla_\varepsilon u(\varepsilon) + \nabla_\varepsilon h(x,\varepsilon)^T \nabla_\varepsilon w(\varepsilon)$$

$$+ \nabla^2_\varepsilon L(x,u,w,\varepsilon) \Big|_{[x(\varepsilon),u(\varepsilon),w(\varepsilon),\varepsilon]} \qquad (7.6)$$

$$= \nabla^2_\varepsilon L(x,u,w,\varepsilon) - N(\varepsilon)^T M(\varepsilon)^{-1} N(\varepsilon) \Big|_{[x(\varepsilon),u(\varepsilon),w(\varepsilon),\varepsilon]} .$$

The next result partially extends Theorem 7.2.

<u>Theorem 7.3</u> (Jittorntrum [12,13]).

If the assumptions of Theorem 4.1 are satisfied, i.e., if the KKT, SSOSC and LI conditions hold at x^* for $P(\varepsilon^*)$ then, in a neighborhood of $\varepsilon = \varepsilon^*$, the local optimal value function f^*_ℓ is once continuously differentiable and twice directionally differentiable:

(a) $f^*_\ell(\varepsilon) = L[x(\varepsilon),u(\varepsilon),w(\varepsilon),\varepsilon]$; (7.7)

$$\text{(b)}\quad \nabla_\varepsilon f^*_\ell(\varepsilon) = \nabla_\varepsilon L(x,u,w,\varepsilon) \Big|_{[x(\varepsilon),u(\varepsilon),w(\varepsilon),\varepsilon]}$$

$$= \nabla_\varepsilon f(x,\varepsilon) - u(\varepsilon)^T \nabla_\varepsilon g(x,\varepsilon)$$

$$+ w(\varepsilon)^T \nabla_\varepsilon h(x,\varepsilon) \Big|_{[x(\varepsilon),u(\varepsilon),w(\varepsilon),\varepsilon]} ; \qquad (7.8)$$

$$\text{(c)}\quad D^2_z f^*_\ell(\varepsilon) = z^T D_z [\nabla_\varepsilon f^*_\ell(\varepsilon)]^T$$

$$= z^T \nabla^2_{y\varepsilon} L D_z y + z^T \nabla^2_\varepsilon L z \Big|_{[x(\varepsilon),u(\varepsilon),w(\varepsilon),\varepsilon]} \qquad (7.9)$$

<u>Proof:</u>

Parts (a) and (b) were obtained by Jittorntrum [12,13]. Part (c) was observed by Fiacco [8] and follows from (b) and Theorem 4.2.

We shall now obtain a further result under the assumptions of Theorem 5.3 as a consequence of a recent more general result due to Rockafellar [26]. Before stating it we need the following definitions:

$$S^*(\varepsilon) = \{x \in R(\varepsilon) \mid f(x,\varepsilon) = f^*(\varepsilon)\} ,$$

$$K_0(x,\varepsilon) = \{(u,w) \mid u_i g_i(x,\varepsilon) = 0,\ u_i \geqslant 0,\ i = 1,\ldots,m,$$

$$- \sum_{i=1}^{m} u_i \nabla_x g_i(x,\varepsilon) + \sum_{j=1}^{p} w_j \nabla_x h_j(x,\varepsilon) = 0\} ,$$

$$Y_S(x,\varepsilon) = \{(u,w) \in K(x,\varepsilon) \mid \text{SOSC holds at } x \text{ with } (u,w) \text{ for } P(\varepsilon)\} .$$

Note that for $x \in R(\varepsilon)$, MFCQ holds at x iff $K_0(x,\varepsilon) = \{0\}$. The function f^* is said to have one-sided directional derivatives at ε in the "Hadamard sense" if the ordinary one-sided directional derivatives

$$D_v f^*(\varepsilon) = \lim_{t \downarrow 0} \frac{1}{t} [f^*(\varepsilon+tv) - f^*(\varepsilon)]$$

exist and in addition

$$D_v f^*(\varepsilon) = \lim_{\substack{v' \to v \\ t \downarrow 0}} \frac{1}{t} [f^*(\varepsilon+tv') - f^*(\varepsilon)] .$$

Theorem 7.4 (Rockafellar [26]).

Let $f^*(\varepsilon^*)$ be finite and an appropriate boundedness assumption be satisfied (see Rockafellar [26]). Suppose that every optimal solution $x \in S^*(\varepsilon^*)$ satisfies $K_0(x,\varepsilon^*) = \{0\}$ and has

$$\text{ri}K(x,\varepsilon^*) \subset Y_S(x,\varepsilon^*) \tag{7.10}$$

where riA denotes the relative interior of A. Then f^* possesses finite one-sided directional derivatives at ε^* in the Hadamard sense, and for every direction v

$$D_v f^*(\varepsilon^*) = \min_{x \in S^*(\varepsilon^*)} \max_{(u,w) \in K(x,\varepsilon^*)} \nabla_\varepsilon L(x,u,w,\varepsilon^*) v . \tag{7.11}$$

Theorem 7.5.

Suppose that the assumptions of Theorem 5.3 are satisfied, i.e., that the KKT, GSSOSC and MFCQ conditions hold at x^* for $P(\varepsilon^*)$. Then, in a neighborhood of $\varepsilon = \varepsilon^*$, the local optimal value function f_ℓ^* has finite one-sided directional derivatives in the Hadamard sense, and for every direction v

$$D_v f_\ell^*(\varepsilon) = \max_{(u,w) \in K(x(\varepsilon),\varepsilon)} \nabla_\varepsilon L(x(\varepsilon),u,w,\varepsilon) v \tag{7.12}$$

Proof:

Under our assumptions it follows from Proposition 7.1 that for some closed ball B around x^* and all ε in some neighborhood N_0

of ε^* , $f_B^*(\varepsilon) = f[x(\varepsilon),\varepsilon]$, where $x(\varepsilon) \in \text{int}B$ is a unique global minimum of $P_B(\varepsilon)$. The boundedness assumption of [26] is obviously satisfied for $P_B(\varepsilon)$. Since the KKT, GSSOSC and MFCQ conditions hold at $x(\varepsilon)$ for $P(\varepsilon)$ (see Theorem 5.3), they also hold for $P_B(\varepsilon)$ by the above. GSSOSC implies that $K(x(\varepsilon),\varepsilon) = Y_S(x(\varepsilon),\varepsilon)$ and MFCQ implies that $K_0(x(\varepsilon),\varepsilon) = \{0\}$. Thus the conclusions of Theorem 7.4 hold for f_B^* , hence also for f_ℓ^* , yielding (7.12).

A weaker version of Theorem 7.5 with $D_v f_\ell^*(\varepsilon)$ being the ordinary one-sided directional derivatives of f_ℓ^* appears in Edahl [4].

It does not seem to be possible to obtain results concerning differentiability of f_ℓ^* under even weaker assumptions than KKT, GSSOSC and MFCQ. This can be seen from Example 6.1, exhibiting nonuniqueness of the local minima (in an arbitrarily small neighborhood of x^*) for perturbed problems under the KKT, SOSC and LI conditions. However, one can still obtain results concerning the restricted optimal value function f_B^* . Under the KKT, SOSC and MFCQ conditions it follows from Theorem 5.1 and its proof (see Robinson [25]) that for a suitably small closed ball B around x^* and all ε in some neighborhood N_0 of ε^*

$$f_B^*(\varepsilon) = \min_{x \in LS(\varepsilon) \cap B^0} f(x,\varepsilon) \tag{7.13}$$

where B^0 denotes the interior of B and $LS(\varepsilon)$ is the nonempty set of local minima of $P(\varepsilon)$. Formula (7.13) reduces to $f_B^*(\varepsilon) = f[x(\varepsilon),\varepsilon]$ if $x(\varepsilon)$ is the unique local minimum of $P(\varepsilon)$ in B^0 as in Proposition 7.1. In general, it asserts that the problem $P_B(\varepsilon)$ attains a global (possibly unique) minimum at some $x \in B^0$. If $LS(\varepsilon) \cap B^0$ is a finite set (as is the case in Example 6.1), then formula (7.13) becomes particularly simple. Of course, one always has $f_B^*(\varepsilon^*) = f(x^*,\varepsilon^*)$ since x^* is the unique global minimum of $P_B(\varepsilon^*)$ under KKT and SOSC. The next result easily follows from Theorem 7.4.

Corollary 7.6.

Suppose that the KKT, GSOSC and MFCQ conditions hold at x^* for $P(\varepsilon^*)$. Then, the restricted optimal value function f_B^* given by (7.13) has finite one-sided directional derivatives at ε^* in the Hadamard sense, and for every direction v

$$D_v f_B^*(\varepsilon^*) = \max_{(u,w) \in K(x^*,\varepsilon^*)} \nabla_\varepsilon L(x^*,u,w,\varepsilon^*)v \tag{7.14}$$

Proof:

Similar to the proof of Theorem 7.5, one notes that GSOSC implies that $K(x^*,\varepsilon^*) = Y_S(x^*,\varepsilon^*)$ and MFCQ implies that $K_0(x^*,\varepsilon^*) = \{0\}$ for $P_B(\varepsilon^*)$ since $x^* \varepsilon \operatorname{int}B$. Since the boundedness assumption of [26] is satisfied for $P_B(\varepsilon^*)$, the conclusions of Theorem 7.4 hold for $f_B^*(\varepsilon)$.

We prove one more result for f_B^* using another result due to Rockafellar [26].

Corollary 7.7.

Suppose that the KKT, SOSC and LI conditions hold at x^* for $P(\varepsilon^*)$. Then, the restricted optimal value function f_B^* given by (7.13) is differentiable at ε^* and

$$\nabla_\varepsilon f_B^*(\varepsilon^*) = \nabla_\varepsilon L(x^*,u^*,w^*,\varepsilon^*) \tag{7.15}$$

where (u^*,w^*) is the unique Lagrange multiplier vector for $P(\varepsilon^*)$ at x^*.

Proof:

The result follows from Rockafellar [26, Corollary, p. 14], since by the LI condition $K(x^*,\varepsilon^*) = Y_S(x^*,\varepsilon^*) = \{(u^*,w^*)\}$ and by the MFCQ condition $K_0(x^*,\varepsilon^*) = \{0\}$ for $P_B(\varepsilon^*)$ with $x^* \varepsilon \operatorname{int}B$, so the assumptions of Rockafellar's corollary hold.

The last two results were proved under the assumptions used in Section 6 (see Robinson's Theorem 6.3 and Example 6.1, [25]).

We conclude by noting that significant practical applications of nonlinear programming sensitivity analysis results are apparent, e.g., see Fiacco and Ghaemi [9], though these have not been addressed in this paper.

REFERENCES

[1] ARMACOST, R.L. and A.V. FIACCO (1978). Sensitivity analysis for parametric nonlinear programming using penalty methods. In Computers and Mathematical Programming, National Bureau of Standards Special Publication 502, 261-269.

[2] BERGE, G. (1963). Topological Spaces. Macmillan, New York.

[3] BIGELOW, J.H. and N.Z. SHAPIRO (1974). Implicit function theorems for mathematical programming and for systems of inequalities, Math. Programming 6 (2), 141-156.

[4] EDAHL, R. (1982). Sensitivity analysis in nonlinear programming, Ph.D. Dissertation, Carnegie-Mellon University.

[5] FIACCO, A.V. (1976). Sensitivity analysis for nonlinear programming using penalty methods, Math. Programming 10 (3), 287-311.

[6] FIACCO, A.V. (1980). Nonlinear programming sensitivity analysis results using strong second order assumptions. In Numerical Optimization of Dynamic Systems (L.C.W. Dixon and G.P. Szego, eds.). North-Holland, Amsterdam, 327-348.

[7] FIACCO, A.V. (1983). Optimal value continuity and differential stability bounds under the Mangasarian-Fromovitz constraint qualification. In Mathematical Programming with Data Perturbations, Vol. II (A.V. Fiacco, ed.). Marcel Dekker, New York, 65-90.

[8] FIACCO, A.V. (1983). Introduction to Sensitivity and Stability Analysis in Nonlinear Programming. Academic Press, New York.

[9] FIACCO, A.V. and A. GHAEMI (1982). Sensitivity analysis of a nonlinear water pollution control model using an upper Hudson River data base, Operations Research 30 (1), 1-28.

[10] FIACCO, A.V. and G.P. McCORMICK (1968). Nonlinear Programming: Sequential Unconstrained Minimization Techniques. Wiley, New York.

[11] GAUVIN, J. (1977). A necessary and sufficient regularity condition to have bounded multipliers in nonconvex programming, Math. Programming 12, 136-138.

[12] JITTORNTRUM, J. (1978). Sequential algorithms in nonlinear programming, Ph.D. Dissertation, Australian National University, Canberra.

[13] JITTORNTRUM, K. (1981). Solution point differentiability without strict complementarity in nonlinear programming, Math. Programming Study 21, Sensitivity, Stability and Parametric Analysis (A.V. Fiacco, ed.), 1984.

[14] KARUSH, W. (1939). Minima of functions of several variables with inequalities as side conditions, M.S. Thesis, Department of Mathematics, University of Chicago.

[15] KOJIMA, M. (1978). Studies on piecewise-linear approximations of piecewise-C^2 mappings in fixed point and complementarity theory, Mathematics of Operations Research 3, 17-36.

[16] KOJIMA, M. (1980). Strongly stable stationary solutions in nonlinear programs. In Analysis and Computation of Fixed Points (S.M. Robinson, ed.). Academic Press, New York, 93-138.

[17] KUHN, H.W. and A.W. TUCKER (1951). Nonlinear programming. In Proceedings of the 2nd Berkeley Symposium on Mathematical Statistics and Probability (J. Neyman, ed.). University of California Press, Berkeley, 481-493.

[18] LEVITIN, E.S. (1975). On the local perturbation theory of a problem of mathematical programming in a Banach space, Soviet Math. Doklady 16, 1354-1358.

[19] McCORMICK, G.P. (1976). Optimality criteria in nonlinear programming. In SIAM-AMS Proceedings 9. SIAM, Philadelphia, 27-38.

[20] ORTEGA, J.M. and W.C. RHEINBOLDT (1970). Iterative Solution of Nonlinear Equations in Several Variables. Academic Press, New York.

[21] PENNISI, L. (1953). An indirect proof for the problem of Lagrange with differential inequalities as added side conditions, Trans. Amer. Math. Soc. 74, 177-198.

[22] ROBINSON, S.M. (1974). Perturbed Kuhn-Tucker points and rates of convergence for a class of nonlinear programming algorithms, Math. Programming 7 (1), 1-16.
[23] ROBINSON, S.M. (1979). Generalized equations and their solutions, Part I: Basic theory, Math. Programming Study 10, 128-141.
[24] ROBINSON, S.M. (1980). Strongly regular generalized equations, Mathematics of Operations Research 5 (1), 43-62.
[25] ROBINSON, S.M. (1982). Generalized equations and their solutions, Part II: Applications to nonlinear programming, Math. Programming Study 19, 200-221.
[26] ROCKAFELLAR, R.T. (1984). Directional differentiability of the optimal value function in a nonlinear programming problem, Math Programming Study 21, Sensitivity, Stability and Parametric Analysis (A.V. Fiacco, ed.).
[27] SPINGARN, J.E. (1977). Generic conditions for optimality in constrained minimization problems, Ph.D. Dissertation, University of Washington, Seattle.
[28] SPINGARN, J.E. (1980). Fixed and variable constraints in sensitivity analysis, SIAM J. Control Optimiz. 18 (3), 297-310.

OPTIMAL DISTURBANCE ATTENUATION WITH CONTROL WEIGHTING[1]

Bruce A. Francis
Department of Electrical Engineering
University of Waterloo
Waterloo, Ontario
Canada N2L 3G1

Abstract

An H_∞-optimal control problem is treated in the context of discrete-time multi-input/output linear systems. The cost is the maximum, over all disturbances of unit energy, of a weighted sum of the energies of the plant's input and output. The cost is minimized over all causal controllers achieving internal stability of the feedback loop.

1. Introduction

A general theory of feedback design should take into account plant disturbances, sensor noise, and plant uncertainty. Moreover, there should be provision for a trade-off between performance and control effort or controller complexity. There is currently underway a program to develop such a theory based on minimax optimization in the frequency domain.

This H_∞ approach to feedback design was introduced in a seminal paper by Zames [1] and has undergone substantial subsequent development [2-15]. The emphasis in [2-5] is on disturbance attenuation or, equivalently, optimization of the sensitivity function. Since control energy is not penalized, the resulting controllers are improper for continuous-time systems (although they are causal in the discrete-time case [13]). Such controllers can, under certain conditions, be approximated by proper ones, with a corresponding slight degradation of performance.

[1] This research was supported by the Natural Sciences and Engineering Research Council of Canada, Grant No. A1715.

Kwakernaak [7,8] considers a more general cost to be minimized: a weighted sum involving both the sensitivity function and its complement. For suitable weightings, the optimal controllers are proper. Verma and Jonckheere [9] treat a similar problem in the single-input/output case, where control energy is explicitly weighted; they reduce the problem to one amenable to Helton's approach [10] to broadband matching. In [11] Helton shows how several classical design constraints can be reformulated as disc constraints in H_∞; however, the resulting problem is not yet solvable exactly. Finally, Doyle [12] poses a very general control problem, which includes plant uncertainty, and outlines an algorithm for computing optimal controllers. The algorithm involves a binary search for the minimal cost: one guesses the value of the minimal cost, performs a computation to determine if the guess is high or low, and then revises the guess.

In this paper the problem of disturbance attenuation is reconsidered, with control energy included in the cost. This problem, which is the multi-input/output generalization of that of Verma and Jonckheere [9], is less general than Helton's [11] or Doyle's [12], but it is believed to be of independent interest because the minimal cost can be expressed as the norm of a certain operator, an observation made by Verma and Jonckheere; this norm can, in principle, be computed a priori.

Consider the following feedback system, where P, C, and W_i $(i = 1,2,3)$ are transfer matrices representing the plant, controller, and certain filters, respectively.

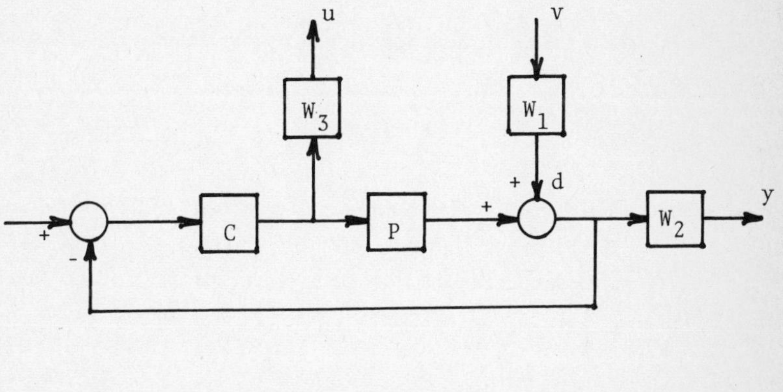

The signal d is the disturbance to be attenuated at the plant output. The class $\mathbb{D}$ of such disturbances is taken to be all outputs of the filter W_1 obtainable from some input v of unit energy:

$$\mathbb{D} = \{d: d = W_1 v, \text{ energy of } v = 1\}.$$

For example, W_1 might be selected so that each disturbance d in $\mathbb{D}$ has its energy concentrated over some frequency band. The signals y and u represent the plant output and input filtered appropriately by W_2 and W_3, respectively. For example, W_3 might be taken to be a high-pass filter in order to penalize high-frequency actuation. Finally, the cost to be minimized is defined as

$$\text{cost} = \sup\,[(\text{energy of } y) + (\text{energy of } u)]^{1/2}, \tag{1}$$

where the supremum is over all d in $\mathbb{D}$.

The design problem is this: P and W_i $(i = 1,2,3)$ are given, and a controller C is to be designed to achieve internal stability of the feedback system and to minimize the cost.

This problem will be solved in the context of discrete-time systems.

2. Notation and Terminology

The set of all complex-valued sequences $\{f_k: -\infty < k < \infty\}$ satisfying $\sum|f_k|^2 < \infty$ is denoted by ℓ_2. Equipped with the inner product

$$<\{f_k\},\{g_k\}>:= \sum \bar{f}_k g_k,$$

ℓ_2 is a Hilbert space. The closed subspace of ℓ_2 of sequences which are zero for $k < 0$ (causal sequences) is denoted by h_2. The orthogonal complement, $h_2^{\perp}$, is therefore the subspace of sequences zero for $k \geq 0$.

Let L_2 denote the class of all complex-valued functions defined on the unit circle $\{e^{j\theta}: -\pi \leq \theta < \pi\}$ and square-integrable with respect to θ. The inner product on L_2 is

$$<f,g>:= (2\pi)^{-1}\int_{-\pi}^{\pi} \overline{f(e^{j\theta})}\,g(e^{j\theta})\,d\theta.$$

The closed subspace of L_2 of functions having analytic continuations into the unit disc is the Hardy space H_2. The theory of Fourier series establishes the isomorphisms

$$L_2 \cong \ell_2, \quad H_2 \cong h_2.$$

For example, the mapping from $\{f_k\}$ in ℓ_2 to f in L_2 is given by

$$f(e^{j\theta}) = \sum f_k e^{jk\theta}.$$

The subspaces of L_2 and H_2 of bounded functions are denoted by L_∞ and H_∞, respectively. The L_∞-norm is

$$\|f\|_\infty := \operatorname*{ess\,sup}_\theta |f(e^{j\theta})|.$$

A prefix R will denote real-rational; for example, RH_∞ is the set of real-rational functions which are analytic in the closed unit disc.

The preceding spaces have vector and matrix versions, denoted by ℓ_2^n, $H_\infty^{n\times m}$, etc. For a matrix F in $L_\infty^{m\times n}$, $\|F\|_\infty$ denotes the essential supremum, over all θ, of the largest singular value of $F(e^{j\theta})$. It can be proved that $\|F\|_\infty$ equals the norm of the operator

$$g \longmapsto Fg, \quad H_2^n \to L_2^m,$$

that is,

$$\|F\|_\infty = \sup \{\|Fg\|_2 : g \in H_2^n, \ \|g\|_2 \le 1\}.$$

The generic complex variable will be denoted by λ. Superscript t on a matrix will denote transpose, and $F^\times(\lambda) := F(\lambda^{-1})^t$. A matrix F in $RH_\infty^{m\times n}$ is <u>inner</u> if $F^\times F = I$ and <u>outer</u> if

$$\text{rank } F(\lambda) = m \quad \text{for all } |\lambda| < 1.$$

The matrix F is said to be <u>*-inner</u> or <u>*-outer</u> if F^t is inner or outer, respectively. Note that for a square matrix, inner is equivalent to *-inner and outer to *-outer. Each matrix F in $RH_\infty^{m\times n}$ has an inner-outer and a *-inner-outer factorization, that is,

$$F = F_i F_o = F_{*o} F_{*i}$$

where F_i is inner, F_o outer, F_{*o} *-outer, and F_{*i} *-inner.

For time-invariant discrete-time linear systems, it is convenient to use λ-transforms, where $\lambda = z^{-1}$. Thus the <u>transfer function</u> of the system

$$y_k = \sum G_{k-i} u_i$$

is taken to be

$$G(\lambda) := \sum \lambda^k G_k.$$

The transfer function $G(\lambda)$ is <u>causal</u> if it is analytic at $\lambda = 0$, <u>strictly causal</u> if, in addition, $G(0) = 0$, and <u>stable</u> if it is analytic in the closed unit disc. All transfer functions in this paper are real-rational.

3. Problem Formulation

<u>It is assumed throughout</u> that P is strictly causal and has no poles on the unit circle and that each W_i is square, stable, and has a stable inverse. Let the dimensions of P be $p\times m$.

The approach in this section is, as usual, to employ the parametrization of Youla et al. [16] of all causal controllers which achieve internal stability, and then to express the cost in terms of the free parameter. To this end, bring in left- and right-coprime factorizations of P: let A, B, $\tilde{A}$, $\tilde{B}$, X, Y, $\tilde{X}$, $\tilde{Y}$ be polynomial matrices satisfying the equations

$$P = \tilde{B}^{-1}\tilde{A} = AB^{-1} \tag{2}$$

$$\tilde{A}\tilde{X} + \tilde{B}\tilde{Y} = I \tag{3}$$

$$XA + YB = I. \tag{4}$$

The formula

$$C = (\tilde{X}+BQ)(\tilde{Y}-AQ)^{-1} \tag{5}$$

$$Q \in RH_\infty^{m\times p}$$

parametrizes all causal real-rational C's which achieve internal stability.

Define the following RH_∞-matrices:

$$T := \begin{bmatrix} W_2\tilde{Y} \\ -W_3\tilde{X} \end{bmatrix}, \quad (p+m)\times p \tag{6}$$

$$U := \begin{bmatrix} W_2A \\ W_3B \end{bmatrix}, \quad (p+m)\times m \tag{7}$$

$$V := \tilde{B}W_1, \qquad p\times p. \tag{8}$$

Lemma 1. With C given by (5), the transfer function from v to $\begin{bmatrix} y \\ u \end{bmatrix}$ equals (T-UQ)V.

Proof. From the figure we have

$$\begin{bmatrix} y \\ u \end{bmatrix} = \begin{bmatrix} W_2(I+PC)^{-1}W_1 \\ -W_3C(I+PC)^{-1}W_1 \end{bmatrix} v, \tag{9}$$

and from (2) and (5) we have

$$\begin{aligned}(I+PC)^{-1} &= [I+\tilde{B}^{-1}\tilde{A}(\tilde{X}+BQ)(\tilde{Y}-AQ)^{-1}]^{-1} \\ &= (\tilde{Y}-AQ)[\tilde{B}(\tilde{Y}-AQ)+\tilde{A}(\tilde{X}+BQ)]^{-1}\tilde{B}.\end{aligned}$$

But the latter quantity in square brackets equals I in view of (2) and (3). Thus

$$(I+PC)^{-1} = (\tilde{Y}-AQ)\tilde{B}.$$

Use of this in (9) yields

$$\begin{bmatrix} y \\ u \end{bmatrix} = (T-UQ)Vv. \quad \blacksquare$$

The cost, as defined in (1), equals

$$\sup\{\| \begin{bmatrix} y \\ u \end{bmatrix} \|_2 : \ v \in H_2^p , \ \|v\|_2 = 1\},$$

and this in turn equals the H_∞-norm of the transfer function from v to $\begin{bmatrix} y \\ u \end{bmatrix}$. For the controller given by (5), we get from Lemma 1 that

$$\text{cost} = \|(T-UQ)V\|_\infty .$$

The objective is thus to minimize $\|(T-UQ)V\|_\infty$ over all stable real-rational matrices Q. A minimizing Q then yields an optimal C via formula (5). Let us therefore define the infimal cost,

$$\nu := \inf\{\|(T-UQ)V\|_\infty : Q \in RH_\infty^{m\times p}\}. \tag{10}$$

4. Existence of an Optimal Controller

It is convenient to manipulate the optimization problem (10) into an alternative form.

The matrices T, U, and V are stable and real-rational. Consider the inner-outer and *-inner-outer factorizations

$$U = U_i U_o , \qquad V = V_{*o} V_{*i} .$$

In view of the specific form (7) of U, it is straightforward to verify that U_o is square and has a stable inverse; this follows from the right-coprimeness of A and B. Similarly, V_{*o} is square and has a stable inverse; this follows from the assumption that P has no poles on the unit circle. Defining

$$Q_1 := U_o Q V_{*o} , \tag{11}$$

we conclude that the mapping $Q \longmapsto Q_1$ is bijective on $RH_\infty^{m\times p}$. Thus from (1) and (11) we get

$$\nu = \inf\{\|(TV_{*o} - U_i Q_1)V_{*i}\|_\infty : Q_1 \in RH_\infty^{m\times p}\}. \tag{12}$$

But since V_{*i} is square and inner, (12) implies that

$$\nu = \inf\{\|TV_{*o} - U_i Q_1\|_\infty : Q_1 \in RH_\infty^{m\times p}\}. \tag{13}$$

The optimization problem in (13) was solved in [4,5] for the special case of square U_i. For the problem at hand, however, U_i is $(p+m)\times m$. Doyle [12] shows that (13) can be transformed into

$$\nu = \inf \left\{ \left\| \begin{bmatrix} R-Q_1 \\ S \end{bmatrix} \right\|_\infty : Q_1 \in RH_\infty^{m\times p} \right\} , \tag{14}$$

where $R \in RL_\infty^{m\times p}$ and $S \in RL_\infty^{p\times p}$. This transformation is based on the following result.

Lemma 2. (Doyle) There exists a matrix U_1 in $RH_\infty^{(p+m)\times p}$ such that $[U_i, U_1]$ is square and inner.

Proof: The matrix $[U_i, U_1]$ is inner if and only if U_1 is inner and $U_i^\times U_1 = 0$. Now U_i is real-rational, stable, and it has rank m (over the field of real-rational functions). Let N be a matrix in $RH_\infty^{(p+m)\times p}$ of rank p and satisfying the equation $U^t N = 0$. (The existence of such N can be justified by consideration of the Smith form of U over the Euclidean domain RH_∞.) By clearing fractions, we can take N to be a polynomial matrix. Then

$$U^\times(\lambda)N(\lambda^{-1}) = 0.$$

Since $N(\lambda^{-1})$ is a polynomial matrix in λ^{-1}, $\lambda^k N(\lambda^{-1})$ is a polynomial matrix in λ for large enough k. Finally, take U_1 to be the inner factor of $\lambda^k N(\lambda^{-1})$. ■

In view of this lemma, we get from (13) that

$$\nu = \inf\{\|TV_{*0} - [U_i, U_1]\begin{bmatrix} Q_1 \\ 0 \end{bmatrix}\|_\infty : Q_1 \in RH_\infty^{m\times p}\}.$$

Thus, to obtain (14), simply define

$$\begin{bmatrix} R \\ S \end{bmatrix} := [U_i, U_1]^{-1} TV_{*0}. \qquad (15)$$

It is now possible to show that the infimum in (14) is achieved and that the infimum equals the norm of a certain operator. Let $H_2^{m\perp}$ denote the orthogonal complement of H_2^m in L_2^m and let Π denote the orthogonal projection

$$L_2^m \longrightarrow H_2^{m\perp}.$$

Define the operator

$$\Gamma : H_2^p \longrightarrow H_2^{m\perp} \times L_2^p$$

$$\Gamma f = \begin{bmatrix} \Pi R f \\ S f \end{bmatrix}. \qquad (16)$$

Theorem. The infimum in (14) is achieved and it equals $\|\Gamma\|$.

The theorem can be proved in several ways: the Nagy-Foias lifting theorem [17] can be used as in [13], or Parrott's theorem [18] can be used as in [19]; the details are omitted.

5. Computation of an Optimal Controller

An optimal C is determined by the following steps: first, the norm of Γ is computed; next, a matrix Q_1 is determined to achieve the infimum in (14); then, Q is determined from (11); finally, C is given by (5).

The norm of Γ equals the square root of the norm of the self-adjoint operator $\Gamma^*\Gamma$ on H_2^p. A matrix representation of $\Gamma^*\Gamma$ can be determined as follows. Each vector f in H_2^p has a power series expansion,

$$f(\lambda) = \sum_0^\infty \lambda^i f_i, \quad f_i \in \mathbb{C}^p.$$

Moreover, $\lambda^i f_i$ is orthogonal to $\lambda^k f_k$ $(i \neq k)$ in the H_2-norm. Therefore we may regard H_2^p as the orthogonal direct sum of its subspace $\lambda^i \mathbb{C}^p$:

$$H_2^p = \mathbb{C}^p \oplus \lambda\mathbb{C}^p \oplus \lambda^2\mathbb{C}^p \oplus \ldots \quad . \tag{17}$$

Take the standard coordinate basis for $\mathbb{C}^p$. Then, corresponding to the decomposition (17), the operator $\Gamma^*\Gamma$ has a block-matrix representation

$$(G_{ij};\ i \geq 0,\ j \geq 0),$$

where G_{ij} is a $p\times p$ matrix which represents that part of $\Gamma^*\Gamma$ mapping $\lambda^j\mathbb{C}^p$ to $\lambda^i\mathbb{C}^p$.

Bring in the power series expansions of the RL_∞-matrices R and S:

$$R(\lambda) = \sum_{-\infty}^{\infty} \lambda^i R_i, \quad S(\lambda) = \sum_{-\infty}^{\infty} \lambda^i S_i.$$

Define the Hankel matrix H_{R^-} corresponding to the $H_2^\perp$-component of R,

$$H_{R^-} := \begin{bmatrix} R_{-1} & R_{-2} & R_{-3} & \cdots \\ R_{-2} & R_{-3} & R_{-4} & \cdots \\ \vdots & \vdots & \vdots & \end{bmatrix}$$

and, finally, define the Hankel matrix H_S corresponding to S,

$$H_S: = \begin{bmatrix} \vdots & \vdots & \vdots & \\ S_1 & S_0 & S_{-1} & \cdots \\ S_0 & S_{-1} & S_{-2} & \cdots \\ S_{-1} & S_{-2} & S_{-3} & \cdots \\ \vdots & \vdots & \vdots & \end{bmatrix}.$$

<u>Lemma 3.</u> <u>The matrix representation of</u> $\Gamma^*\Gamma$ <u>is</u>

$$H^t_{\tilde R} H_{\tilde R} + H^t_S H_S.$$

<u>Proof</u>. Let $\lambda^i f$ and $\lambda^j g$ denote generic elements of $\lambda^i \mathbb{C}^p$ and $\lambda^j \mathbb{C}^p$, respectively, i.e., $f,g \in \mathbb{C}^p$. Then, by definition, G_{ij} satisfies

$$\langle \lambda^i f, \Gamma^*\Gamma\lambda^j g\rangle = f^* G_{ij} g.$$

Using the definition (16) of Γ, we have

$$\begin{aligned}\langle \lambda^i f, \Gamma^*\Gamma\lambda^j g\rangle &= \langle \Gamma\lambda^i f, \Gamma\lambda^j g\rangle \\ &= \langle \Pi R\lambda^i f, \Pi R\lambda^j g\rangle + \langle S\lambda^i f, S\lambda^j g\rangle .\end{aligned}$$

Since

$$\Pi R\lambda^i f = \sum_{k=-\infty}^{-1} \lambda^k R_{k-i} f,$$

we get that

$$\langle \Pi R\lambda^i f, \Pi R\lambda^j g\rangle = f^* \Big(\sum_{k=-\infty}^{-1} R^t_{k-i} R_{k-j}\Big) g.$$

Similarly,

$$\langle S\lambda^i f, S\lambda^j g\rangle = f^* \Big(\sum_{k=-\infty}^{\infty} S^t_{k-i} S_{k-j}\Big) g.$$

Thus

$$G_{ij} = \sum_{k=-\infty}^{-1} R^t_{k-i} R_{k-j} + \sum_{k=-\infty}^{\infty} S^t_{k-i} S_{k-j}. \blacksquare$$

Let

$$\Pi_n : H_2^p \longrightarrow H_2^p$$

denote the orthogonal projection onto the first n components in the right-hand side of (17). Then

$$\lim_{n\to\infty} \|\Pi_n \Gamma^* \Gamma \Pi_n\| = \|\Gamma^*\Gamma\| .$$

Thus the norm of $\Gamma^*\Gamma$ can be computed to any desired accuracy by computing the norm (largest eigenvalue) of the finite matrix

$$(G_{ij};\quad 0 \le i,\quad j \le n)$$

for sufficiently large n.

Assume that $\|\Gamma^*\Gamma\|$, hence ν, has been computed. Then an optimal Q_1 can be determined as follows[2]. It is apparent from (14) that $\nu \ge \|S\|_\infty$.

Case 1. $\nu > \|S\|_\infty$.

The procedure is to reduce the problem to one which has already been solved. It is convenient to introduce the following notation: if F and G are square matrices with elements in RL_∞, then $F \ge G$ means that the matrix

$$F(e^{j\theta})-G(e^{j\theta})$$

is positive semi-definite for all θ. It follows that the inequalities

$$\|S\|_\infty \le \nu \qquad S^\times S \le \nu^2 I$$

are equivalent.

Now fix Q_1 in $RH_\infty^{m\times p}$. From (14), Q_1 is optimal if and only if

$$\left\| \begin{bmatrix} R - Q_1 \\ S \end{bmatrix} \right\|_\infty = \nu \ . \tag{18}$$

This implies that

$$(R-Q_1)^\times(R-Q_1) + S^\times S \le \nu^2 I. \tag{19}$$

The matrix $\nu^2 I - S^\times S$ is positive definite on the unit circle because $\nu > \|S\|_\infty$. Thus there exists a matrix M in $RH_\infty^{p\times p}$ such that M^{-1} is stable and

$$\nu^2 I - S^\times S = M^\times M.$$

Then (19) is equivalent to the condition

$$\|RM^{-1}-Q_2\|_\infty \le 1, \tag{20}$$

where $Q_2 := Q_1 M^{-1}$. Actually, only equality can hold in (20), since strict inequality would contradict the equality in (18). We conclude that Q_1

[2] This procedure was developed independently by Doyle [12] and the author [13].

is optimal if and only if the distance from Q_2 to RM^{-1} is minimal, i.e., one.

In summary, for Case 1 an optimal Q_1 can be determined from R and S as follows:

Step 1. Compute $\nu = \|\Gamma^*\Gamma\|^{1/2}$.

Step 2. Find a real-rational stable M such that M^{-1} is stable and $\nu^2 I - S^\times S = M^\times M$.

Step 3. Find Q_2 in $RH_\infty^{m\times p}$ to minimize

$$\|RM^{-1} - Q_2\|_\infty.$$

Step 4. Set $Q_1 = Q_2 M$.

Step 2 is spectral factorization. Step 3, finding an RH_∞-matrix closest to a given RL_∞-matrix, can be done using the Nevanlinna-Pick theory, as in [5], the Ball-Helton theory, as in [4,12,20], or some other method.

Case 2. $\nu = \|S\|_\infty$.

In this case the matrix $\nu^2 I - S^\times S$ is only positive semi-definite on the unit circle; the corresponding spectral factor M is outer, but M^{-1} has poles on the unit circle. Thus the method employed in Case 1 does not apply.

In lieu of a general theory for Case 2, it is proposed that a Q_1 be computed which is only suboptimal, but arbitrarily close to optimal. To do this, choose ν_1 slightly greater than $\|S\|_\infty$ and carry out Steps 2 to 4 above using ν_1 in place of ν.

6. Concluding Remarks

In the feedback loop considered in this paper, there was one input, v, and two outputs, y and u. The cost was taken as the H_∞-norm of the transfer function from v to $\begin{bmatrix} y \\ u \end{bmatrix}$.

A natural generalization is to consider an additional input to the feedback loop, for example a reference input r. An appropriate cost may be the H_∞-norm of the transfer function from $\begin{bmatrix} r \\ v \end{bmatrix}$ to $\begin{bmatrix} e \\ u \end{bmatrix}$, where e is the tracking error. It can be shown that the transfer function takes the form

$$T - UQV,$$

where T, U, V are stable matrices, U having more rows than columns and V having more columns than rows. A complete theory for the problem

$$\underset{Q\ \text{stable}}{\text{minimize}} \|T-UQV\|_\infty$$

remains to be developed.

References

[1] G. Zames, Feedback and optimal sensitivity, IEEE Trans. Aut. Control, AC-26, 1981, 301-320.

[2] G. Zames and B.A. Francis, Feedback, minimax sensitivity and optimal robustness, ibid, AC-28, 1983, 585-601.

[3] B.A. Francis and G. Zames, On H^∞-optimal sensitivity theory for siso feedback systems, ibid, AC-29, 1984, 9-16.

[4] B.A. Francis, J.W. Helton, and G. Zames, H^∞-optimal feedback controllers for linear multivariable systems, Proc. Symp. Math. Theory of Networks and Systems, Israel, 1983; to appear IEEE Trans. Aut. Control.

[5] B-C. Chang and J.B. Pearson, Optimal disturbance reduction in linear multivariable systems, to appear IEEE Trans. Aut. Control.

[6] M.G. Safonov and M.S. Verma, L^∞-sensitivity optimization via Hankel-norm approximation, Proc. Amer. Control Conf., San Francisco, 1983.

[7] H. Kwakernaak, Minimax frequency domain performance and robustness optimization of linear feedback systems, Tech. Report, Dept. Appl. Math., Twente Univ. of Tech., 1983.

[8] H. Kwakernaak, Minimax frequency domain optimization of multivariable linear feedback systems, Proc. IFAC World Congress, Budapest, 1984.

[9] M.S. Verma and E. Jonckheere, L^∞-compensation with mixed sensitivity as a broadband matching problem, submitted for publication, 1984.

[10] J.W. Helton, Broadbanding: gain equalization directly from data, IEEE Trans. Circuits and Systems, CAS-28, 1981, 1125-1137.

[11] J.W. Helton, Worst case analysis in the frequency domain: the H^∞ approach to control, Proc. IEEE Conf. Dec. Control. San Antonio, 1983.

[12] J. Doyle, Synthesis of robust controllers and filters, ibid.

[13] B.A. Francis, Notes on H^∞-optimal linear feedback systems, Lectures presented at Linköping University, Aug. 22-26, 1983.

[14] P.P. Khargonekar and A. Tannenbaum, Noneuclidean metrics and the robust stabilization of systems with parameter uncertainty, Tech. Report, Center for Math. System Theory, Univ. of Florida, Feb. 1984.

[15] Y.K. Foo and I. Postlethwaite, An H^∞-minimax approach to the design of robust control systems, OUEL Report No. 1520/84, Dept. of Eng. Sci., Univ. of Oxford, Mar. 1984.

[16] D.C. Youla, H. Jabr, and J. Bongiorno, Jr., Modern Wiener-Hopf design of optimal controllers, Part II: the multivariable case, IEEE Trans. Aut. Control, AC-21, 1976, 319-338.

[17] B. Sz.-Nagy and C. Foias, Harmonic Analysis of Operators on Hilbert Space, American Elsevier, N.Y., 1970.

[18] S. Parrott, On a quotient norm and the Sz.-Nagy-Foias lifting theorem, Journal of Functional Analysis, 30, 1978, 311-328.

[19] S.C. Power, Hankel operators on Hilbert space, Pitman, Boston, 1982.

[20] B.A. Francis and G. Zames, Design of H^∞-optimal multivariable feedback systems, Proc. IEEE Conf. Dec. Control, San Antonio, 1983.

PATH-FOLLOWING METHODS FOR KUHN-TUCKER CURVES BY AN ACTIVE INDEX SET STRATEGY*

H.Gfrerer, Hj.Wacker, W.Zulehner
Institut für Mathematik
Johannes Kepler Universität Linz
A-4040 Linz, Austria

J.Guddat
Sektion Mathematik
Humboldt-Universität Berlin
DDR-108 Berlin, German Democratic Republic

Abstract

In this paper Kuhn-Tucker curves of one-parameter optimization problems $P(t)$, $t \in \mathbb{R}$ are discussed. Using an active index set strategy, path following procedures for $P(t)$ are developed. One-parameter optimization problems naturally arise if a nonlinear optimization problem P is solved by a homotopy method. Under certain conditions on P it is shown that the proposed homotopy method is constructive with probability one in the sense of Chow, Mallet-Paret, Yorke [4].

Key words: Path-following methods, Kuhn-Tucker curves, one-parameter optimization problems, homotopy method, active index set strategy.

* Research partially supported by the Austrian Fonds zur Förderung der wissenschaftlichen Forschung, contract 4786.

1. Introduction

In this paper we discuss one-parameter optimization problems of the form

$$P(t) \quad \min_{x \in \mathbb{R}^n} h_o(x,t)$$
$$\text{s.t. } h_j(x,t) = 0, \; j \in L = \{1,\dots,l\} \tag{1.1}$$
$$h_j(x,t) \le 0, \; j \in M = \{l+1,\dots,m\},$$

depending on a parameter $t \in \mathbb{R}$, where $h_j: \mathbb{R}^{n+1} \to \mathbb{R}$, $j = 0,1,\dots,m$, are twice continuously differentiable. Precisely, we study stationary solutions of P(t), i.e. solutions $(x,u) \in \mathbb{R}^{n+m}$ of the corresponding Kuhn-Tucker system:

$$KT(t) \quad \nabla_x h_o(x,t) + \sum_{j \in L \cup M} u_j \nabla_x h_j(x,t) = 0$$
$$h_j(x,t) = 0, \; j \in L$$
$$h_j(x,t) \le 0, \; u_j \ge 0, \; u_j h_j(x,t) = 0, \; j \in M,$$

if the parameter t changes. Following Kojima, Hirabayashi [15] these conditions can be formulated as a system of equations:

$$H(x,y,t) = 0, \tag{1.2}$$

where $H: \mathbb{R}^{n+m+1} \to \mathbb{R}^{n+m}$ is defined by

$$H(x,y,t) = \begin{bmatrix} \nabla_x h_o(x,t) + \sum_{j \in L} y_j \nabla_x h_j(x,t) + \sum_{j \in M} y_j^+ \nabla_x h_j(x,t) \\ -h_j(x,t), \; j \in L \\ y_j^- - h_j(x,t), \; j \in M \end{bmatrix}$$

with $y_j^+ = \max(y_j,0), y_j^- = \min(y_j,0)$.

Obviously, if (x,u) solves KT(t), then $H(x,y,t) = 0$ with $y_j = u_j$ for $j \in L$, $y_j = u_j + h_j(x,t)$ for $j \in M$, and every solution (x,y,t) of (1.2) leads to a solution (x,u) of KT(t) with $u_j = y_j$ for $j \in L$,

$u_j = y_j^+$ for $j \in M$ and $h_j(x,t) = y_j^-$ for $j \in M$.

The mapping H is piecewise continuously differentiable with respect to the subdivision $\{\tau(J) | J \subset M\}$ of $\mathbb{R}^{n+m+1}$, where the cells $\tau(J)$ are defined by

$$\tau(J) = \mathbb{R}^n \times \{y \in \mathbb{R}^m | y_j \geq 0 \text{ for } j \in J,\ y_j \leq 0 \text{ for } j \in M\setminus J\} \times \mathbb{R},$$

i.e. for each cell $\tau(J)$ there exist an open set $U \supset \tau(J)$ and a continuously differentiable function $G: U \to \mathbb{R}^{n+m}$ with $G|_{\tau(J)} = H|_{\tau(J)}$.

The basic assumption in this paper is that 0 is regular value of H, i.e. for each J, $J_o \subset M$, there is an open set $U \supset \sigma(J,J_o)$ and a continuously differentiable function $G: U \to \mathbb{R}^{n+m}$, such that $G|_{\sigma(J,J_o)} = H|_{\sigma(J,J_o)}$ and the Jacobian of $G \circ i$ is of rank n+m at each point z with $i(z) \in U$ and $G(i(z)) = 0$, where $\sigma(J,J_o) = \{(x,y,t) \in \tau(J) | y_j = 0 \text{ for } j \in J_o\}$ and i is the natural embedding from $\mathbb{R}^n \times \mathbb{R}^{M\setminus J_o} \times \mathbb{R}$ into $\mathbb{R}^{n+m+1}$.

<u>Remark:</u> For $J_o \neq \emptyset$, $\sigma(J,J_o)$ is said to be a face of $\tau(J)$.

Then it can be shown (see Kojima, Hirabayashi [15], Theorem 2.1, (4.9)) that

(a) each connected component S of $H^{-1}(0) = \{(x,y,t) \in \mathbb{R}^{n+m+1} | H(x,y,t) = 0\}$ is either homeomorphic to $\mathbb{R}$ (path) or to the unit circle $\{(s_1,s_2) \in \mathbb{R}^2 | s_1^2 + s_2^2 = 1\}$ (loop) and, for each $\tau(J)$, $J \subset M$, with $S \cap \tau(J) \neq \emptyset$, each connected component of $S \cap \tau(J)$ is diffeomorphic to an interval or to the unit circle (piecewise continuously differentiable)

(b) (transversality condition) if S intersects with a face σ of some cell $\tau(J)$, $J \subset M$, at (x,y,t), then there is a unique index $j_o \in M$ with $y_{j_o} = 0$, and S traverses σ at (x,y,t), i.e. the one-sided tangent vectors at (x,y,t) are not parallel to σ;

(c) if the connected component S is a path, i.e. $S = \{\theta(s) | s \in \mathbb{R}\}$, then $\|\theta(s)\| \to \infty$ if $s \to \pm \infty$.

For details see Kojima, Hirabayashi [15].

Many studies have been made on existence and regularity of Kuhn-Tucker curves, see e.g. Dontchev, Jongen [5], Kojima [14], Kojima, Hirabayashi [15], Robinson [17]. This paper is mainly based on results in [15]. The structure of one-parameter optimization problems is utilized in homotopy methods for solving nonlinear optimization problems. Several computational methods were developed, see e.g. Allgower, Georg [1] for a general survey, Burkhardt, Richter [3], Eaves [6], Eaves, Scarf [7], Garcia, Gould [8], Garcia, Zangwill [9], Hackl [12], Kojima [13], Todd [2o], Watson [21], Zangwill, Garcia [22].

In this paper $P(t)$ is approached by an active index set strategy. For quadratic programming, see Ritter [16], Tammer [19]. In preceding papers [1o], [11] the authors applied this technique to homotopy methods for nonlinear optimization problems under the assumption that for the homotopy $P(t)$ there exist solutions $(x(t), u(t))$, $t \in [0,1]$, satisfying regularity conditions. In this paper the existence of a solution path can be shown if the original problem satisfies certain assumptions.

In Section 2 stability sets and critical points are introduced for $P(t)$ and their properties are shown. This allows to reduce $P(t)$ to an optimization problem with equality constraints. These theoretical results are utilized to develop path-following methods for $P(t)$ in Section 3. Finally, in Section 4 a particular homotopy $\bar{P}(t)$ for a nonlinear optimization problem P is proposed, for which the existence of a path with probability one is shown.

2. Theoretical background

Let S be a connected component of $H^{-1}(0)$. If 0 is regular value of H, then the results in Section 1 imply that each compact connected subset of S can be described parametrically in the form

$\{\theta(s)=(x(s),\ y(s),\ t(s)) \in \mathbb{R}^{n+m+1} | s \in I\}$, I compact interval, for some continuous function θ.

Let $J_o(x,t)$ be the index set of active inequalities of $P(t)$ at $(x,t) \in \mathbb{R}^{n+1}$, i.e. $J_o(x,t)=\{j\in M \mid h_j(x,t)=0\}$. In particular, $J_o(x(s), t(s))$ is denoted by $J_o(s)$. Then we have

Theorem 2.1

Assume that 0 is regular value of H. Then there exist a unique subdivision of I into a finite number of nontrivial intervals

$$I = \bigcup_{k=1}^{K} [s^{k-1}, s^k]$$

and a corresponding family of index sets J^k, $k=1,\dots,K$, such that

(1) $J^k \neq J^{k+1}$, $k=1,\dots,K-1$

(2) $\theta(]s^{k-1}, s^k[) \subset \operatorname{int} \tau(J^k)$

(3) $J_o(s) = J^k$ for all $s \in]s^{k-1}, s^k[$.

Proof:

Consider $A = \bigcup_{J \subset M} \theta^{-1}(\partial\tau(J)) \cup \partial I$.

If A is finite, i.e. $A = \{s^o, s^1, \dots s^K\}$ with $s^o < s^1 < \dots < s^K$, then, for each $k\in\{1,\dots,K\}$, there exists a unique index set $J^k \subset M$, such that $\theta(]s^{k-1}, s^k[) \subset \operatorname{int} \tau(J^k)$. Hence, for each $s \in]s^{k-1}, s^k[$, we have $y_j(s) > 0$ for $j \in J^k$ and $y_j(s) < 0$ for $j \in M\setminus J^k$. Since $h_j(x(s), t(s)) = y_j(s)^-$, it follows that $J_o(s) = J^k$. The transversality condition guarantees that the curves $\theta(s)$ traverses the boundary $\partial\tau(J^k)$ at the point $\theta(s^k)$, $k=1,\dots,K-1$, which gives (1).

Now assume that A is infinite. Since I is compact, there exists a sequence $\{s^i\}$ in $\bigcup_{J \subset M} \theta^{-1}(\partial\tau(J))$ with $s^i \to \bar{s} \in \bigcup_{J \subset M} \theta^{-1}(\partial\tau(J))$, $s^i \neq \bar{s}$. The transversality condition says that the curve $\theta(s)$ traverses the boundary of a cell at $\theta(\bar{s})$. Hence, for some neighborhood U of $\bar{s}$, $\theta(U) \cap \bigcup_{J \subset M} \partial\tau(J) = \{\bar{s}\}$, which contradicts the existence of the sequence $\{s^i\}$.

The uniqueness is obvious.

□

The intervals $[s^{k-1},s^k]$ are said to be stability sets, the points s^k, $k=1,\dots,K-1$, are called critical points. Different stability sets may correspond to the same active index set J^k. Property (1) of Theorem 1 excludes only the equality of index sets corresponding to successive stability sets.

On stability sets Theorem 2.1 allows to replace problem P(t) by a reduced problem of the form

$$P^J(t) \quad \min_{x\in\mathbb{R}^n} \; h_o(x,t)$$
$$\text{s.t. } h_j(x,t) = 0, \; j \in L \cup J$$

with $J \subset M$. The corresponding Kuhn-Tucker system is

$$H^J(x,u,t) = 0 \tag{2.1}$$

where $H^J: \mathbb{R}^n \times \mathbb{R}^{L\cup J} \times \mathbb{R} \to \mathbb{R}^n \times \mathbb{R}^{L\cup J}$ is defined by

$$H^J(x,u,t) = \begin{bmatrix} \nabla_x h_o(x,t) + \sum\limits_{j\in L\cup J} u_j \nabla_x h_j(x,t) \\ h_j(x,t), \; j \in L \cup J \end{bmatrix}$$

The next theorem gives a computationally important characterization of stability sets:

<u>Theorem 2.2</u>

Assume the notations and hypotheses of Theorem 2.1.
For each $k\in\{1,\dots,K\}$, there exict an open neighborhood $U^k \supset [s^{k-1},s^k]$ and a continuously differentiable function $\theta^k: U^k \to \mathbb{R}^n \times \mathbb{R}^{L\cup J^k} \times \mathbb{R}$, $\theta^k(s) = (x^k(s), u^k(s), t^k(s))$, with

(1) $x^k(s) = x(s)$, $u_j^k(s) = y_j(s)$ for $j \in L \cup J^k$, $t^k(s) = t(s)$ for all $s \in [s^{k-1},s^k]$,

(2) $\theta^k(s)$ solves (2.1) with $J = J^k$ for all $s \in U^k$.

(3) $[s_{k-1}, s_k]$ is a connected component of

$$\{s \in U^k \cap I \mid \ h_j(x^k(s), t^k(s)) \leq 0,\ j \in M\setminus J^k \text{ and } u_j^k(s) \geq 0,\ j \in J^k\}$$

Proof:

Since 0 is regular value of H, the matrix

$$\left[\begin{array}{c|c|c|c}
\nabla_{xx}h_o + \sum\limits_{j\in L\cup J^k} y_j \nabla_{xx}h_j & \nabla_x h_j^T & 0 & \nabla_{xt}h_o^T + \sum\limits_{j\in L\cup J^k} y_j \nabla_{xt} h_j^T \\ \hline
-\nabla_x h_j & 0 & 0 & -\nabla_t h_j \\ \hline
-\nabla_x h_j & 0 & E & -\nabla_t h_j
\end{array}\right]
\begin{array}{l} \\ \\ \}\ j\in L\cup J^k \\ \\ \}\ j\in M\setminus J^k \end{array}$$

$$\qquad\qquad\qquad\quad \underbrace{\qquad}_{j\in L\cup J^k}\ \underbrace{\qquad}_{j\in M\setminus J^k}$$

(E identity matrix) has full rank n+m at (x(s), y(s), t(s)) for each $k\in\{1,\dots,K\}$ and each $s \in [s^{k-1}, s^k]$. This implies that the Jacobian of H^{J^k}, which is obtained by deleting the rows and columns with indices $j \in M\setminus J^k$, is of full rank at the corresponding points. Then the implicit function theorem gives (1) and (2).

Let $A := \{s\in U^k \cap I \mid \ h_j(x^k(s), t^k(s)) \leq 0,\ j\in M\setminus J^k$ and

$u_j^k(s) \geq 0,\ j\in J^k\}$

We have $[s^{k-1}, s^k] \subset A$ by (1).

Assume that there exists a connected set B with $[s^{k-1}, s^k] \subset B \subset A$. Then, without loss of generality, $]s^k, s^k+\varepsilon[\ \subset B$ and $]s^k, s^k+\varepsilon[\ \subset [s^k, s^{k+1}]$ for some $\varepsilon > 0$. (The case that $]s^{k-1}-\varepsilon, s^{k-1}[\ \subset B$, $]s^{k-1}-\varepsilon, s^{k-1}[\ \subset [s^{k-2}, s^{k-1}]$ can be treated analogously.)

Since $J^k \neq J^{k+1}$, there is an index $j \in M$ with either $j \in J^k\setminus J^{k+1}$ or $j \in J^{k+1}\setminus J^k$. If $j \in J^k\setminus J^{k+1}$, then $y_j(s) = u_j^k(s) > 0$ for $s \in \ ]s^{k-1}, s^k[$, $u_j^k(s^k) = 0$, and, by the transversality condition, $\dot u_j^k(s^k) = \lim\limits_{s\to s^k-} \dot y_j(s) \neq 0$. Hence $u_j^k(s) < 0$ for $s \in \ ]s^k, s^k+\varepsilon[$,

ε sufficiently small. This implies $]s^k, s^k + \varepsilon[\cap A = \emptyset$ in contradiction to the assumption.

If $j \in J^{k+1} \setminus J^k$, then $y_j(s) = h_j(x^k(s), t^k(s)) < 0$ for $s \in]s^{k-1}, s^k[$,

$h_j(x^k(s^k), t^k(s^k)) = 0$, and, by the transversality condition,

$\frac{d}{ds} h_j(x^k(s), t^k(s))\Big|_{s=s^k} = \lim_{s \to s^k-} \dot{y}_j(s) \neq 0$. Hence $h_j(x^k(s), t^k(s)) > 0$

for $s \in]s^k, s^k + \varepsilon[$, ε sufficiently small. This implies $]s^k, s^k+\varepsilon[\cap A = \emptyset$, which contradicts the assumption.

□

The last theorem of this section describes how the index set J^{k+1} varies from the index set J^k:

Theorem 2.3

Assume the notations and hypotheses of Theorem 2.2

For each $k \in \{1, \ldots, K-1\}$, there is a unique index $j^k \in M$ with either $u^k_{j^k}(s^k) = 0$ and $j^k \in J^k$ or $h_{j^k}(x^k(s^k), t^k(s^k)) = 0$ and $j^k \in M \setminus J^k$, and

$$J^{k+1} = \begin{cases} J^k \cup \{j^k\} & \text{if } j^k \in M \setminus J^k \\ J^k \setminus \{j^k\} & \text{if } j^k \in J^k \end{cases}$$

Proof:

Since $\theta(s^k) \in \partial\tau(J^k)$ for $k \in \{1, \ldots K-1\}$, there is a unique index $j^k \in M$ with $y_{j^k}(s^k) = 0$ by the transversality condition. That means that, for $j \in J^k$, we have $u^k_{j^k}(s^k) = 0$ and, for $j^k \in M \setminus J^k$ $h_{j^k}(x^k(s^k), t^k(s^k)) = 0$. The rest follows from Theorem 2.2 (3).

□

3. The algorithm

3.1. Standard path following methods for nonlinear equations

On a stability set $[s^{k-1}, s^k]$ Problem P(t) reduces to a system of equations $H^J(x,u,t) = 0$ with $J = J^k$. Algorithms that numerically follow a path of solutions of such a system are well-known, see e.g. Allgower, Georg [1]. We restrict ourselves to predictor-corrector methods.

Assume that z^{p-1} approximates a point $z(s^{p-1})$ of the path of solutions $z(s)$, $s \in I$, I interval, of the system

$$F(z) = 0,$$

$F: \mathbb{R}^{N+1} \to \mathbb{R}^N$ continuously differentiable, and that A^{p-1} is an approximation of the Jacobian $\nabla F(z^{p-1})$.
Let $\pi \in \{-1,+1\}$. Then the p-th iteration step of the method consists of two parts:

(a) predictor step
From z^{p-1} we take a first step along a direction w^{p-1} with step length σ^{p-1}

$$z^{p,0} = z^{p-1} + \sigma^{p-1} w^{p-1},$$

where w^{p-1} solves

$$A^{p-1} . w = 0, \quad \|w\| = 1, \text{ sign det} \begin{bmatrix} w^T \\ A^{p-1} \end{bmatrix} = \pi .$$

(w^{p-1} is an approximation of the unit tangent vector of the path of solutions at $z(s^{p-1})$. The value of π is kept constant during the iterations.)

(b) corrector step
Then we use $z^{p,0}$ as a starting point of a corrector iteration:

$$A^{p,q-1} \cdot (z^{p,q} - z^{p,q-1}) + F(z^{p,q-1}) = 0$$
$$(w^{p,q-1})^T (z^{p,q} - z^{p,q-1}) = 0 \qquad q=1,\ldots,q_p$$

where $A^{p,q-1}$ approximates $\nabla F(z^{p,q-1})$. Usually, $w^{p,q-1}$ is equal to w^{p-1} or to one of the natural basis vectors of $\mathbb{R}^{N+1}$. We set

$$z^p = z^{p,q_p},$$

which is an approximation of a solution $z(s^p)$.

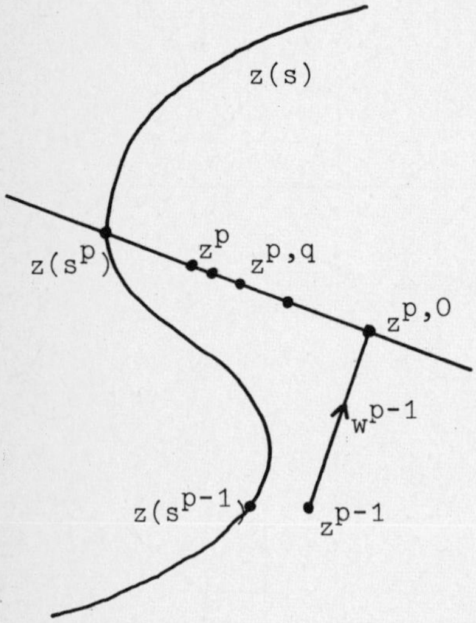

In this figure it is assumed that $w^{p,q-1} = w^{p-1}$, $q = 1,\ldots,q_p$.

This general class of predictor-corrector methods includes the Euler-predictor ($A^{p-1} = \nabla F(z^{p-1})$), the Newton-corrector ($A^{p,q-1} = \nabla F(z^{p,q-1})$, $w^{p,q-1} = w^{p-1}$), and also Quasi-Newton techniques.

3.2. Path-following for P(t)

If a critical point s^k is passed, the index set J of the nonlinear system $H^J(x,u,t) = 0$ which is equivalent to P(t) changes from J^k to J^{k+1}. So, for P(t), the standard path-following techniques have to be extended by a procedure for determining the index set after passing critical points.

We propose the following algorithm:

Let $\pi^{p-1} \in \{-1,+1\}$ and $J^{p-1} \subset M$. Assume that $z^{p-1} = (x^{p-1},u^{p-1},t^{p-1}) \in \mathbb{R}^n \times \mathbb{R}^{L\cup J^{p-1}} \times \mathbb{R}$ approximates a solution of $H^{J^{p-1}}(x,u,t) = 0$ with $h_j(x^{p-1},t^{p-1}) \leq 0$ for all $j \in M\setminus J^{p-1}$ and $u_j^{p-1} \geq 0$ for all $j \in J^{p-1}$.

Then the p-th iteration step is of the form:

(a) predictor step for the system $H^{J^{p-1}}(z) = 0$, $z = (x,u,t)$

$$z^{p,0} = z^{p-1} + \sigma^{p-1}.\ w^{p-1},$$

where w^{p-1} solves

$$A^{p-1}w = 0,\ \|w\| = 1,\ \text{sign det}\begin{bmatrix} w^T \\ A^{p-1} \end{bmatrix} = \pi^{p-1}.$$

(b) corrector step for the system $H^{J^{p-1}}(z) = 0$:

$$\left.\begin{aligned} &A^{p,q-1}(z^{p,q} - z^{p,q-1}) + H^{J^{p-1}}(z^{p,q-1}) = 0 \\ &(w^{p,q-1})^T (z^{p,q} - z^{p,q-1}) = 0 \end{aligned}\right. \qquad q = 1,\ldots,q_p$$

Set $\bar{z}^p = (\bar{x}^p, \bar{u}^p, \bar{t}^p) = z^{p,q_p}$

(c) Check whether the stability set was left:
If $h_j(\bar{x}^p, \bar{t}^p) \leq 0$ for all $j \in M\setminus J^{p-1}$ and $\bar{u}_j^p \geq 0$ for all $j \in J^{p-1}$, set:

$$z^p = \bar{z}^p,\ J^p = J^{p-1},\ \pi^p = \pi^{p-1}.$$

(d) Otherwise, there is an index $j^p \in M$ with either $h_{j^p}(\bar{x}^p, \bar{t}^p) > 0$ and $j^p \in M\setminus J^{p-1}$ or $\bar{u}^p_{j^p} < 0$ and $j^p \in J^{p-1}$.

Then change the index set:

$$J^p = \begin{cases} J^{p-1} \cup \{j^p\} & \text{if } j^p \in M\setminus J^{p-1} \\ J^{p-1} \setminus \{j^p\} & \phantom{\text{if }} j^p \in J^{p-1} \end{cases}$$

and set

$$z^p = (\bar{x}^p, u^p, \bar{t}^p) \in \mathbb{R}^n \times \mathbb{R}^{L \cup J^p} \times \mathbb{R} \text{ with}$$

$$u_j^p = \begin{cases} \bar{u}_j^p & \text{for } j \in J^p \cap J^{p-1} \\ 0 & j \in J^p \setminus J^{p-1} \end{cases}$$

and

$$\pi^p = -\pi^{p-1}.$$

Theorem 2.2 suggests step (c), while step (d) is partly motivated by Theorem 2.3. The change of the sign of π is suggested by

Lemma 3.1

Assume the notations and the hypotheses of the theorems in Section 2. Then, for $k = \{1,\dots,K-1\}$, we have

$$\lim_{s \to s^k_-} \text{sign det} \begin{bmatrix} \dot{\theta}^k(s)^T \\ \hline \nabla H^{J^k}(\theta^k(s)) \end{bmatrix} = - \lim_{s \to s^k_+} \text{sign det} \begin{bmatrix} \dot{\theta}^{k+1}(s)^T \\ \hline \nabla H^{J^{k+1}}(\theta^{k+1}(s)) \end{bmatrix}$$

Proof:

In Garcia, Zangwill [9] 2.3 it is shown that

$$\dot{\theta}_j(s) = c(s).\ \det \begin{bmatrix} e_j^T \\ \hline \nabla H(\theta(s)) \end{bmatrix} \quad \text{for all } s \in I \setminus \{s^0, s^1, \dots, s^K\},$$

where sign $c(s)$ is constant with respect to s and e_j is the j-th natural basis vector in $\mathbb{R}^{n+m+1}$.

Let $k \in \{1,\dots,K\}$ and $s \in (s^{k-1}, s^k)$. Then, by deleting all columns with index $j \in M \setminus J^k$ and the corresponding rows, we obtain

$$\det \begin{bmatrix} e_j^T \\ \hline \nabla H(\theta(s)) \end{bmatrix} = (-1)^{|M \setminus J^k|} \det \begin{bmatrix} e_j^T \\ \hline \nabla H^{J^k}(\theta^k(s)) \end{bmatrix}$$

(e_j denotes the j-th natural basis vector in $\mathbb{R}^{n+m+1}$ resp. $\mathbb{R}^m \times \mathbb{R}^{L \cup J^k} \times \mathbb{R}$.)

Therefore,

$$\det \begin{bmatrix} \dot{\theta}^k(s)^T \\ \hline \nabla H^{J^k}(\theta^k(s)) \end{bmatrix} = \sum_j \dot{\theta}^k_j(s) \, . \det \begin{bmatrix} e_j^T \\ \hline \nabla H^{J^k}(\theta^k(s)) \end{bmatrix} =$$

$$= c(s).(-1)^{|M \setminus J^k|} . \sum_j \det \begin{bmatrix} e_j^T \\ \hline \nabla H^{J^k}(\theta^k(s)) \end{bmatrix}^2$$

Since the cardinalities of J^k and J^{k+1} differ by 1, the proof is completed.

□

For the path-following method with Euler-predictor and Newton-corrector the numerical feasibility can be shown by standard arguments, i.e. any compact connected subset of a connected component S of $H^{-1}(0)$ can be numerically followed with arbitrary accuracy, as long as the initial approximation is sufficiently good and the step lengths are sufficiently small.

The same holds for appropriate Quasi-Newton variants of this method.

4. Application: Globally convergent algorithms for nonlinear optimization problems

One-parameter optimization problems arise quite naturally if an ordinary optimization problem

$$\begin{aligned} P: \quad & \min_{x \in \mathbb{R}^n} \; f_o(x) \\ & \text{s.t.} \; f_j(x) = 0, \; j \in L \\ & \qquad f_j(x) \le 0, \; j \in M \end{aligned}$$

is solved by a homotopy method: P is embedded into a family of problems $P(t)$, $t \in \mathbb{R}$, of the form (1.1), such that $P(0)$ has a known solution (x^o,u^o) and $P(1) = P$. The basic idea of the method is to follow the solution path of $P(t)$ as t changes from 0 to 1, starting at the known solution (x^o,u^o) for $t = 0$ and finally ending up with a solution of P for $t = 1$.

If 0 is regular value of H (see (1.2)), then the connected component S_o of $H^{-1}(0)$ which contains $(x^o,y^o,0)$ is a piecewise continuously differentiable path or loop. Feasibility of the homotopy method additionally requires that S_o reaches the hyperplane $t = 1$.

Recall that $P(t)$ satisfies the Mangasarian-Fromovitz constraint qualification (M-F) if, for each

$x \in X(t) = \{x \in \mathbb{R}^n | h_j(x,t) = 0 \text{ for } j \in L,\ h_j(x,t) \le 0 \text{ for } j \in M\}$

(a) the set $\{\nabla_x h(x,t) | j \in L\}$ is linearly independent

(b) there exists an vector $z \in \mathbb{R}^n$ such that

$$\nabla_x h_j(x,t)z = 0 \quad \text{for } j \in L$$
$$\nabla_x h_j(x,t)z < 0 \quad \text{for } j \in J_o(x,t)$$

Theorem 4.1

Assume that

(1) 0 is regular value of H
(2) for each t in some neighborhood of 0, there is a unique solution (x^t,y^t) of $H(.,.,t) = 0$
(3) $\emptyset \neq X(t) \subset C$, C compact, for each $t \in [0,1]$
(4) $P(t)$ satisfies M-F for each $t \in [0,1]$

Then S_o is a piecewise continuously differentiable path which connects $(x^o,y^o,0)$ with some point $(x^*,y^*,1)$.

Proof:

(1) and (2) imply that there exists a path $\theta: \mathbb{R} \to \mathbb{R}^{n+m+1}$, $\theta(s)=(x(s),y(s),t(s))$ with $\theta(0)=(x^o,y^o,0)$ and $t(s) > 0$ for each $s > 0$.

Assume that $t(s) < 1$ for each $s > 0$. Since $\|\theta(s)\| \to \infty$ for $s \to \infty$ and $t(s) \in (0,1)$, $x(s) \in C$ for $s > 0$, there is a sequence $\{s^i\}$ with $s^i \to \infty$, $t(s^i) \to \bar{t} \in [0,1]$, $x(s^i) \to \bar{x}$ and $\|y(s^i)\| \to \infty$. Because $y_j(s^i)^- = h_j(x(s^i), t(s^i)) \to h_j(\bar{x},\bar{t})$ for $j \in M$, the sequence $\|u(s^i)\|$ tends to infinity, too.
(4) implies that the point-to-set mapping $U(x,t) = \{u | (x,u)$ is solution of $KT(t)\}$ is upper semicontinuous at $(\bar{x},\bar{t})$ and $U(\bar{x},\bar{t})$ is bounded ([18], Theorem 2.3), which contradicts $\|u(s^i)\| \to \infty$.

□

For Problem P we assume that

(A1) $\emptyset \neq X = \{x \in \mathbb{R}^n | f_j(x) = 0$ for $j \in L$, $f_j(x) \leq 0$ for $j \in M\}$ and X is compact,

(A2) for each stationary solution (x,u) of $P(t)$

(a) the set $\{\nabla_x f_j(x) | j \in L \cup J_o(x)\}$ is linearly independent with $J_o(x) = \{j \in M | f_j(x) = 0\}$

(b) the strict complementary slackness condition holds for (x,u)

(c) $v^T \nabla_{xx} L(x,u) v \neq 0$ for each $v \neq 0$ with

$$\nabla_x f_j(x).v = 0 \text{ for } j \in L \cup J_o(x)$$

(L is the Lagrangian of P)

The first assumption implies that there is a convex differentiable function $f_{m+1}: \mathbb{R}^n \to \mathbb{R}$, such that

$X \subset C(r) = \{x \in \mathbb{R}^n | f_{m+1}(x) \leq r\}$ with $C(r)$ compact and int $C(r) \neq \emptyset$ for all $r \in [r_1,r_2]$, $r_1 < r_2$ (e.g. $f_{m+1}(x) = \|x\|^2$).

Next we assume that P, extended by the constraint $f_{m+1}(x) = r$, satisfies a constraint qualification stronger than M-F:

(A3) For each $r \in [r_1,r_2]$ and for each $x \in C(r)$

(a) the set $\{\nabla_x f_j(x) | j \in L\}$ is linearly independent

(b) there is a vector $z \in \mathbb{R}^n$ with

$$f_j(x) + \nabla_x f_j(x).z = 0 \quad \text{for } j \in L$$

$$f_j(x) + \nabla_x f_j(x).z < 0 \quad \text{for } j \in M \text{ with } f_j(x) \geq 0$$

$$\nabla_x f_{m+1}(x).z < 0 \text{ if } f_{m+1}(x) = r$$

Finally,

(A4) f_i, $i=0,\ldots,m+1$ are three times continuously differentiable.

The particular embedding

$$\bar{P}(t): \quad \min_{\bar{x} \in \mathbb{R}^{\bar{n}}} h_o(\bar{x},t)$$

$$\text{s.t.} \quad h_j(\bar{x},t) = 0, \; j \in L = \{1,\ldots,l\}$$

$$h_j(\bar{x},t) \leq 0, \; j \in \bar{M} = \{l+1,\ldots,\bar{m}\}$$

that we study in this section is defined by

$$h_o(\bar{x},t) = t(f_o(x) + \tfrac{1}{2}\|v\|^2) + (1-t).\tfrac{1}{2}\,(|x-x^o|^2 + \|v-v^o\|^2)$$

$$h_j(\bar{x},t) = t\, f_j(x) + (1-t)\,(v_j-d_j), \; j \in L \cup M$$

$$h_{m+1}(\bar{x},t) = f_{m+1}(x) - r$$

$$h_{m+2}(\bar{x},t) = \|v\|^2 - q$$

with $\bar{x} = (x,v) \in \mathbb{R}^{\bar{n}}$, $\bar{n} = n+m$, $\bar{M} = M \cup \{m+1,m+2\}$, $\bar{m} = m+2$, for some fixed parameters $w = (x^o,v^o,d,r,q) \in \mathbb{R}^{n+2m} \times [r_1,r_2] \times]0,\infty[$.

The function $\bar{H}(\bar{x},\bar{y},t;w)$ for $\bar{P}(t)$ is analogously defined as $H(x,y,t)$ for $P(t)$ in Section 1. Let $\bar{J}_o(\bar{x},t) = \{j \in \bar{M} | h_j(\bar{x},t) = 0\}$ and $\bar{X}(t) = \{\bar{x} \in \mathbb{R}^{\bar{n}} | h_j(\bar{x},t) = 0 \text{ for } j \in L \text{ and } h_j(\bar{x},t) \leq 0 \text{ for } j \in \bar{M}\}$.

It is easy to see that any solution $\bar{x}^*$ of $\bar{P}(1)$ is of the form $\bar{x}^* = (x^*,0)$ and corresponds to a solution x^* of P, and vice versa. In this sense we have $\bar{P}(1) = P$.

Now we are able to prove

Theorem 4.2

Assume that P satisfies (A1), (A2), (A3), (A4) and let

$$W=\{w=(x^o,v^o,d,r,q) \in \mathbb{R}^{n+2m+2} \mid f_{m+1}(x^o)<r, \|d\|<q,\ v_j^o>d_j>0 \text{ for } j\in M,$$
$$r_1<r\le r_2,\ q>0\}.$$

Then, for almost all $w = (x^o,v^o,d,r,q) \in W$, the point $(\bar{x}^o,\bar{u}^o)$ with $\bar{x}^o = (x^o,d)$, $\bar{u}^o = (v^o-d,0,0)$ is the unique stationary solution of $\bar{P}(0)$, and the connected component $\bar{S}_o$ of $\bar{H}^{-1}(0)$ which contains $(\bar{x}^o,\bar{y}^o,0)$ is a piecewise continuously differentiable path that connects $(\bar{x}^o,\bar{y}^o,0)$ with some point $(\bar{x}^*,\bar{y}^*,1)$, where $\bar{x}^* = (x^*,0)$ and x^* solves P.

Proof:

It is easy to see that $(\bar{x}^o,\bar{u}^o)$ is unique stationary solution of $\bar{P}(0)$ and that $\bar{P}(0)$ is strongly regularin the sense of Robinson [17], see Theorem 4.1 in [17]. Then, by [17] Theorem 2.1, $\bar{P}(t)$ has a unique solution for each t in some neighborhood of 0.

$\bar{P}(0)$ satisfies M-F:

Let $\bar{x} =(x,v) \in \bar{X}(0)$. Then $\nabla_{\bar{x}}h_j(\bar{x},0) = e^T_{n+j}$ for $j \in L \cup M$, where e_{n+j} is the $(n+j)$-th natural basis vector in $\mathbb{R}^{n+m}$.

Therefore,

$\{\nabla_{\bar{x}}h_j(\bar{x},0) \mid j \in L\}$ is linearly independent, and for $\bar{z} \in \mathbb{R}^{\bar{n}}$ with

$$\bar{z}_j = x_j^o-x_j \quad \text{for } j \in \{1,\dots,n\}$$

$$\bar{z}_j = \begin{cases} 0 & \text{for } j \in L \\ -v_j & \text{for } j \in M \end{cases}$$

we have

$$\nabla_{\bar{x}}\, h_j(\bar{x},0)\bar{z} = \bar{z}_{n+j} = 0 \quad \text{for } j \in L,$$

$$\nabla_{\bar{x}}h_j(\bar{x},0).\bar{z} = \bar{z}_{n+j} = -v_j = -d_j < 0 \quad \text{for } j \in \bar{J}_o(\bar{x},0) \cap M,$$

$$\nabla_{\bar{x}} h_{m+1}(\bar{x},0)\bar{z} = \nabla_x f_{m+1}(x)(x^o - x) \le f_{m+1}(x^o) - f_{m+1}(x) = f_{m+1}(x^o) - r < 0$$

$$\text{if } f_{m+1}(x) = r,$$

$\nabla_{\bar{x}} h_{m+2}(\bar{x},0)\bar{z} = -2 \sum_{j \in M} v_j^2 < 0$ if $\|v\|^2 = q > 0$, because

$\|v\|^2 = q$ and $v_j = 0$ for $j \in M$ imply $q = \sum_{j \in L} v_j^2 = \sum_{j \in L} d_j^2 \le \|d\|^2$

in contradiction to $\|d\|^2 < q$.

For $t \in]0,1[$ $\bar{P}(t)$ satisfies M-F, too:

Let $\bar{x} = (x,v) \in \bar{X}(t)$. Then $\nabla_{\bar{x}} h_j(\bar{x},t) = (t\nabla_x f_j(x),(1-t)e_{n+j}^T)$ for $j \in L \cup M$.

Therefore,

$\{\nabla_{\bar{x}} h_j(\bar{x},t) \mid j \in L\}$ is linearly independent.

By (A3) b, there is a vector $z \in \mathbb{R}^n$ with

$f_j(x) + \nabla_x f_j(x)z = 0$ for $j \in L$,

$f_j(x) + \nabla_x f_j(x)z < 0$ for $j \in M$ with $f_j(x) \ge 0$,

$\nabla_x f_{m+1}(x)z < 0$ if $f_{m+1}(x) = r$.

Obviously, for $j \in M^- = \{j \in \bar{J}_o(\bar{x},t) \cap M \mid f_j(x) < 0\}$ there is a number $\lambda_j > 1$ with

$\lambda_j f_j(x) + \nabla_x f_j(x)z < 0$

and $v_j - d_j = -\frac{t}{1-t} f_j(x) > 0$, hence $v_j > d_j > 0$.

Then, for $\bar{z} \in \mathbb{R}^{\bar{n}}$, given by

$$\bar{z}_j = z_j \qquad \text{for } j \in \{1,\dots,n\}$$

$$\bar{z}_{j+n} = \begin{cases} d_j - v_j & \text{for } j \in L \cup (M \setminus M^-) \\ \lambda_j(d_j - v_j) & \text{for } j \in M^- \end{cases}$$

we have

$$\nabla_{\bar{x}} h_j(\bar{x},t)\bar{z} = t\nabla_x f_j(x)z + (1-t)(d_j - v_j) = -tf_j(x) - (1-t)(v_j - d_j) = 0 \text{ for } j \in L$$

$$\nabla_{\bar{x}} h_j(\bar{x},t)\bar{z} = t\nabla_x f_j(x)z + (1-t)(d_j - v_j) < -tf_j(x) - (1-t)(v_j - d_j) = 0$$

$$\text{for } j \in \bar{J}_o(\bar{x},t) \cap M,\ f_j(x) \ge 0,$$

$$\nabla_{\bar{x}} h_j(\bar{x},t)\bar{z} = t\nabla_x f_j(x)z+(1-t)\lambda_j(d_j-v_j) < -t.\lambda_j f_j(x)-(1-t)\lambda_j(v_j-d_j) = 0$$

$$\text{for } j \in \bar{J}_0(\bar{x},t) \cap M,\ f_j(x) < 0,$$

$$\nabla_{\bar{x}} h_{m+1}(\bar{x},t)\bar{z} = \nabla_x f_{m+1}(x).z < 0 \text{ if } f_{m+1}(x) = r,$$

$$\nabla_{\bar{x}} h_{m+1}(\bar{x},t)\bar{z} = 2 \sum_{j\in L\cup M} v_j(d_j-v_j) + 2 \sum_{j\in M^-} \underbrace{v_j(\lambda_j-1)(d_j-v_j)}_{<\,0}$$

$$\le \|d\|^2 - \|v\|^2$$

$$= \|d\|^2 - q < 0 \quad \text{if} \quad \|v\|^2 = q$$

(A3) directly implies M-F for $\bar{P}(1)$.

Of course, $\bar{X}(t) \subset \bar{C}(r,q) = C(r) \times \{v \in \mathbb{R}^n \mid \|v\|^2 \le q\}$ and $\bar{C}\,(r,q)$ is compact.

It remains to show that 0 is regular value of $\bar{H}$ for almost all $w \in W$. Since

$$\bar{H}(\bar{x},\bar{y},t;w) = \bar{H}(\bar{x},\bar{y},t;0) + (-(1-t)x^o,-(1-t)v^o,(1-t)d,r,q)^T,$$

this follows from Sard's Lemma ((A2) guarantees full rank at t=1), see Kojima [14], remark after Theorem 7.1.

Therefore, by applying Theorem 4.1 the proof is completed.

□

Remark:

The embedding $\bar{P}(t)$ is motivated by classical penalty methods: For simplicity assume that $x^o = v^o = d^o = 0$ and that the last two constraints are ignored. Then the auxiliary variables v_j can be directly computed for $t < 1$:

$$v_j = -\frac{t}{1-t} \cdot f_j(x) \qquad \text{for } j \in L$$

$$v_j = -\frac{t}{1-t} \cdot \max(f_j(x),0) \qquad \text{for } j \in M$$

and we obtain for the objective function

$$h_o(x,t)=tf_o(x)+\frac{1}{2}\left(\frac{t}{1-t}\right)^2\left(\sum_{j\in L} f_j(x)^2+\sum_{j\in M}\max(f_j(x),0)^2\right)+\frac{1}{2}(1-t)\|x\|^2,$$

which is typical for penalty methods. Observe that the expression

$\frac{1}{2}(1-t)\|x\|^2$ can be interpreted as a regularization term.

References

[1] E.Allgower and K.Georg, "Simplicial and continuation methods for approximating fixed points and solutions to systems of equations", SIAM Review 22 (198o) 28-85.

[2] B.Bank, J.Guddat, B.Kummer, D.Klatte and K.Tammer, Nonlinear parametric optimization (Akademie-Verlag Berlin, 1982).

[3] S.Burkhardt, C.Richter, "Ein Prädiktor-Korrektor-Verfahren der nichtlinearen Optimierung", Wissenschaftliche Zeitschrift der Technischen Universität Dresden 31 (1982) 193-198.

[4] S.N.Chow, J.Mallet-Paret and J.A.Yorke, "Finding zeroes of maps: homotopy methods that are constructive with probability one", Mathematics of Computations 32 (1978) 887-899.

[5] A.L.Dontchev, H.Th.Jongen, "On the regularity of the Kuhn-Tucker curves", Memorandum NR. 46o, Department of Applied Mathematics, Twente University of Technology (Enschede, The Netherlands, 1984)

[6] B.C.Eaves, "A short course in solving equations with PL homotopies", in: R.W.Cottle and C.E.Lemke, eds., Nonlinear programming, Proceedings of the Ninth SIAM-AMS Symposium in Applied Mathematics, (SIAM, Philadelphia, 1976) pp. 73-143.

[7] B.C.Eaves and H.E.Scarf, "The solution of systems of piecewise linear equations", Mathematics of Operations Research 1 (1976) 1-27.

[8] C.B.Garcia and F.J.Gould, "An application of homotopy to solving linear programs", Mathematical Programming 27 (1983) 263-282.

[9] C.B.Garcia and W.I.Zangwill, Pathways to solutions, fixed points and equilibria (Prentice-Hall, Inc., Englewood Cliffs, 1981).

[1o] H.Gfrerer, J.Guddat and Hj.Wacker, "A globally convergent algorithm based on imbedding and parametric optimization", Computing 3o (1983) 225-252.

[11] J.Guddat, Hj.Wacker and W.Zulehner, "On imbedding and parametric optimization - a concept of a globally convergent algorithm for nonlinear optimization problems", Mathematical Programming Study, to appear.

[12] J.Hackl, "Solution of optimization problems with nonlinear restrictions via continuation methods", in: Hj.Wacker, ed., Continuation methods (Academic Press, New York, 1978) pp. 95-127.

[13] M.Kojima, "A complementary pivoting approach to parametric nonlinear programming", Mathematics of Operations Research 4 (1979) 464-477.

[14] M.Kojima, "Strongly stable stationary solutions in nonlinear programming", in: S.M.Robinson, ed., Analysis and computation of fixed points (Academic Press, 198o) pp. 93-138.

[15] M.Kojima and R.Hirabayashi, "Continuous deformations of nonlinear programs", Technical Report 1o1, Department of Information Sciences, Tokyo Institute of Technology (Tokyo, 1981).

[16] K.Ritter, "Ein Verfahren zur Lösung parameterabhängiger nichtlinearer Maximumprobleme", Unternehmungsforschung 6 (1962) 149-196.

[17] S.M.Robinson, "Strongly regular generalized equations", Mathematics of Operations Research 5 (198o) 43-62.

[18] S.M.Robinson, "Generalized equations and their solutions, part II: applications to nonlinear programming", Mathematical Programming Study 19 (1982) 2oo-221.

[19] K.Tammer, "Die Abhängigkeit eines quadratischen Optimierungsproblems von einem Parameter in der Zielfunktion", Mathematische Operationsforschung und Statistik 5 (1974) 573-59o.

[2o] M.J.Todd, "A quadratically-convergent fixed-point algorithm for economic equilibria and linearly constrained optimization", Mathematical Programming 18 (198o) 111-126.

[21] L.T.Watson, "Solving the nonlinear complementary problem by a homotopy method", SIAM Journal on Control and Optimization 17 (1979) 36-46.

[22] W.I.Zangwill and C.B.Garcia, "Equilibrium programming: the path following approach and dynamics", Mathematical Programming 21 (1981) 262-289.

Two case-studies in parametric semi-infinite programming

Prof. Dr. R. Hettich
Universität Trier
FB IV, Postfach 3825
55-Trier (Germany)

Dr. P. Zencke
Goethestr. 27
6901-Dossenheim (Germany)

Abstract. It is shown that the problem of approximating the eigenvalues and eigenfunctions of a homogeneous membrane as well as that of approximating (in the sense of Chebyshev) a function by generalized rational functions can be treated as semi-infinite programming problems depending on a real parameter.

In this contribution we discuss the applicability of parametric programming to two well-known problems of applied analysis. The first one is that of computing eigenvalues and eigenfunctions of the Δ-operator for regions in the plane. The underlying physical problem is the determination of eigenvalues and -functions of an oscillating, homogeneous membrane which has applications in different fields of engineering. We will show that approximate eigenvalues and eigenfunctions can be computed by choosing a paramter $\lambda\in\mathbb{R}$ in a certain parametric semi-infinite programming problem in such a way that the value of this problem is minimized locally. At present algorithms based on this method are not sufficiently developed to decide whether it may compete with the more established methods of finite differences or finite elements. However, examples indicate that at least in special cases the new method performs excellently and even remarkably better than the others. So further research in this field seems appropriate.

The second problem to be discussed is that of approximating in the sense of Chebshev a uni- or multivariate function by generalized rationals. Again this problem can be formulated as a parametric semi-infinite programming problem, where now the parameter $\lambda\in\mathbb{R}$ has to be adapted in such a way that the value of this problem becomes zero. Contrary to the first problem in this case we may conclude definitely that the resulting algorithms perform excellently. They are closely related to the original version of the differential correction algorithm [2], which is considered the most reliable for generalized rational approximation. However, in addition, the proposed parametric programming approach enables a general and satisfactory theory of convergence as well as a numerical implementation which is very effective with regard to storage requirements and computing time [12].

In Section 1 we present the eigenvalue problem, its formulation as a parametric programming problem which is based on a theorem of Moler and Payne, and examples of the value of this problem as a function of the parameter, the local minima of which give estimates and bounds for the eigenvalues. Section 2 deals with the rational approximation problem and its parametric programming formulation. In Section 3 we investigate the problem of differentiating the value function $\mu(\lambda)$ of semi-infinite programming problems depending on a parameter $\lambda\in\mathbb{R}$. In the final two sections we apply these results to our problems and give some numerical examples.

1. Approximate computation of eigenvalues of the Laplace-operator

1.A. Statement of problem

In the sequel let $R\subset\mathbb{R}^2$ be a bounded, simply connected, and open region with boundary B. Then we consider the problem of finding a positive $\lambda\in\mathbb{R}$ and $u\in C^2(R)\cap C(\overline{R}), \overline{R}=R\cup B$, $u\neq o$, such that

$$\Delta u(x) + \lambda u(x) = o \quad \text{for } x\in R \tag{1.1.a}$$

$$u(x) = o \quad \text{for } x\in B. \tag{1.1.b}$$

This is one of the classical problems in mathematical physics: Suppose we are given a homogeneous elastic membrane stretched in a plane frame B enclosing the region R. Then, for $x\in R$, $t\geq t_o$, let $U(x,t)$ be the deflection of point x of the membrane at time t. Then, with c a constant depending on mass and elasticity of the membrane, its motion is governed by the partial differential equation

$$\Delta U(x,t) = cU_{tt} \tag{1.2}$$

(where Δ operates on U as a function of x). Scaling time such that c=1, problem (1.1) is obtained from (1.2) when we separate the variables

$$U(x,t) = u(x)\ v(t) \tag{1.3}$$

to find stationary oscillations.

We define an inner product on C(R B) by

$$(f,g) = \frac{1}{|\bar{R}|} \int_{\bar{R}} f(x)g(x)dx \tag{1.4}$$

with $|\bar{R}|$ the area of $\bar{R}$. As usual, a norm is defined by $\|f\| = (f,f)^{1/2}$.

The following theorem is classical (cf. [3]).

Theorem 1. Suppose B is such that Green's function exists for operator Δ. Then there exists a sequence of eigenvalues λ_i for (1.1), $0<\lambda_1<\lambda_2<\ldots$, having no (finite) point of accumulation, and an orthogonal, in the space $\{f\in C^2(R)\cap C(\bar{R})\,|\,f(x)=0,\ x\in B\}$ L^2-complete system of eigenfunctions being complete in the compact-open topology, too.

If λ,u solve (1.1) it is easily seen that

$$U(x,t) = u(x)(\alpha \cos\sqrt{\lambda}\ t + \beta \sin\sqrt{\lambda}\ t) \tag{1.5}$$

solves (1.2), i.e. the deflection u(x) at time $t_o=0$ is modulated in time with frequency $\frac{\sqrt{\lambda}}{2\pi}$.

To find solutions of (1.1.a), we introduce polar coordinates $z=\binom{r}{\theta}$ $r = \sqrt{x_1^2 + x_2^2}$, $\theta = \text{arc tg}\ \frac{x_2}{x_1}$. Then, with $w(r,\theta)=u(x(r,\theta))$ we get from (1.1.a)

$$w_{rr} + \frac{1}{r} w_r + \frac{1}{r^2} w_{\theta\theta} + \lambda w = 0 . \tag{1.6}$$

Separating now the variables r and θ, i.e. taking

$$w(r,\theta) = g(r)h(\theta) \tag{1.7}$$

we get in a straightforward way two ordinary differential equations in r and θ resp.

$$h''(\theta) + \nu^2 h(\theta) = 0 \tag{1.8a}$$

$$r^2 g''(r) + rg'(r) + (r^2\lambda-\nu^2)g(r)=0 \tag{1.8b}$$

with arbitrary $\lambda\in\mathbb{R}$ and some constant ν. Or, conversely, if h and g solve (1.8) for some $\nu\in\mathbb{R}$, then (1.7) solves (1.6) and when returning to the original variables x_1,x_2 we find a solution of (1.1.a).

However, letting $\rho=\sqrt{\lambda}r$, (1.8b) becomes the Bessel-equation such that for every $\nu \in \mathbb{R}$ and $\lambda > 0$ we find solutions

$$\begin{aligned} \phi_\nu^c(\lambda;z) &= J_\nu(\sqrt{\lambda}r)\cos(\nu\theta) \\ \phi_\nu^s(\lambda;z) &= J_\nu(\sqrt{\lambda}r)\sin(\nu\theta) \end{aligned} \tag{1.9}$$

of (1.6) resp. of (1.1.a).

We remark that, given R, sequences $\{\nu_i\}$ may be constructed such that the set $\{\phi_{\nu_i}^c(\lambda;\cdot), \phi_{\nu_i}^s(\lambda;\cdot)\}_{i=1,\ldots}$ is complete in the sense of Theorem 1. Therefore, if λ is an eigenvalue, any eigenfunction to λ may be approximated uniformly in any compact $R' \subset R$, by linear combinations of functions of type (1.9) (cf. [5]).

1.B. An inclusion-theorem of Moler and Payne

The following theorem is a consequence of the maximum-principle for harmonic functions and the completeness of the set of eigenfunctions (cf. Theorem 1). It has been proved for more general problems in [8], cf. also [4].

Theorem 2. Given $\hat\lambda > 0$ und $\hat u \in C^2(R) \cap C(\bar R)$, $\|\hat u\| = 1$, such that

$$\Delta \hat u + \hat\lambda \hat u = 0 \text{ in } R \tag{1.10}$$

and

$$\varepsilon = \max_{x \in B} |\hat u(x)| < 1. \tag{1.11}$$

Then we have

(a) There exists an eigenvalue λ_k of (1) such that

$$\frac{\hat\lambda}{1+\varepsilon} \le \lambda_k \le \frac{\hat\lambda}{1-\varepsilon} \tag{1.12}$$

(b) There exists an eigenfunction u_k to λ_k such that

$$\|\hat u - u_k\| \le \frac{\varepsilon}{\alpha}(1+\varepsilon^2/\alpha^2)^{1/2} \tag{1.13}$$

where $\alpha = \min\{ \frac{|\lambda_1 - \hat\lambda|}{\lambda_1} \mid \lambda_1 \neq \lambda_k\}$.

Quite similar results hold for general self-adjoint operators on a separable Hilbert space with a completely continuous inverse. The crucial point w.r.t. the practical applicability is the construction of $\hat u$ (given $\hat\lambda$) such that (1.10) holds. In our special case by (1.9) infinitely many u are specified to a given $\hat\lambda$. Nickel [9] gives inclusions for the case that (1.10) is not met exactly. Yet, there seems to be no numerical experience based on Nickel's result.

To Problem (1.1) Theorem 2 has been applied in the past as follows ([4,5,8]): A number N of points $x_1,\ldots,x_N \in B$ is chosen together with N functions $\varphi_j(\lambda;x)$ of type (1.9) (depending on λ). Then $\hat{\lambda}$ and parameters $\hat{\alpha}_j$ are determined such that

$$\varphi(\hat{\lambda};x_i) = \varphi_1(\hat{\lambda};x_i) + \sum_{j=2}^{N} \hat{\alpha}_j \varphi_j(\hat{\lambda};x_i) = 0 \quad , \quad i = 1,\ldots, N.$$

Now the Theorem is applied with $\hat{\lambda}$ and $\hat{u} = \varphi(\hat{\lambda};x)/\|\varphi(\hat{\lambda};\cdot)\|$. To avoid the determination of $\|\varphi(\hat{\lambda};\cdot)\|$, requiring an integration over the domain R, lower bounds on $\|\varphi(\hat{\lambda};\cdot)\|$ are computed, yielding an upper bound for ε.

We have chosen another approach which is more flexible and, for more complicated shapes R of the membrane, more efficient too, as it usually requires a lower number of approximating functions to obtain a given accuracy.

1.C. A parametric programming approach to the membrane-problem

Suppose we have specified a sequence $0 \le \nu_1 < \nu_2 < \ldots$. Let (cf.(1.9))

$$\phi_N(\lambda,z) = \sum_{i=1}^{N} [\alpha_i \phi^c_{\nu_i}(\lambda,z) + \beta_i \phi^s_{\nu_i}(\lambda,z)].$$

Then, for all $\lambda > 0$, and $\alpha_i, \beta_i \in \mathbb{R}$

$$\Delta\phi_N(\lambda,z) + \lambda\phi_N(\lambda,z) = 0, \quad z \in R.$$

Therefore, given $\lambda \in \mathbb{R}_+$, we can compute an error bound $\mu(\lambda)$ for λ according to Theorem 2 by solving the problem

<u>SIP (λ):</u> Minimize $\ell(\alpha,\beta,\varepsilon) = \varepsilon$ subject to the constraints

$$\left.\begin{aligned} \phi_N(\lambda,z) &\le \varepsilon \\ -\phi_N(\lambda,z) &\le \varepsilon \end{aligned}\right\} z \in B$$

and some normalization such as $\alpha_j = 1$ for some j appropriately chosen.

Let $\bar{\alpha}$, $\bar{\beta}$, $\bar{\varepsilon} = \varepsilon(\lambda)$ be a solution of SIP(λ), $\bar{\phi}_N(\lambda,z) = \sum_{i=1}^{N} [\bar{\alpha}_i \phi^c_{\nu_i} + \bar{\beta}_i \phi^s_{\nu_i}]$
Then a relative error-bound for λ to an eigenvalue of (1.1) is given by

$$\mu(\lambda) = \varepsilon(\lambda)/\|\bar{\phi}_N(\lambda,\cdot)\|.$$

This suggests to find approximations $\hat{\lambda}$ to eigenvalues of the membrane by determining (local) minima of the function $\mu(\lambda)$.

Note that in order to determine really optimal errorbounds for given $\{\nu_i\}$ and N, instead of the normalization in SIP(λ) one should choose the constraint $\|\phi_N(\lambda,\cdot)\| = 1$. This, however, results in nonlinear problems SIP(λ) requiring an enormous amount of computational effort.

1.D. Examples

We shall illustrate the approach by giving examples of functions $\mu(\lambda)$ for different regions R.

Example 1. Let R be the ellipse with half diameters 2 and 1 resp., i.e.

$$R = \{x \mid x_1^2/4 + x_2^2 < 1\} .$$

It can be shown (cf.[6]) that it is sufficient to consider only a quarter of R and B and to choose in advance four different types of function sequences according to the four possible types of symmetries w.r.t. the axes of the ellipse, each eigenfunction being either symmetric or anti-symmetric to the axes.

For instance, to determine eigenvalues with eigenfunctions symmetric w.r.t. both axes a proper choice is

$$\phi_N^{\alpha}(\lambda,z) = \sum_{i=0}^{N-1} \alpha_i J_{2i}(\sqrt{\lambda}r) \cos 2i\theta$$

For N = 5 we obtain $\mu(\lambda)$ as shown in Figure 1. Table 1 shows that four eigenvalues are approximated rather closely already which indeed are the first four in the sequence of eigenvalues with eigenfunctions symmetric to both axes

$\hat{\lambda}$	$\mu(\hat{\lambda})$	bound	$\lambda-\hat{\lambda}$
3.566 726.583	$0.502 \cdot 10^{-7}$	$0.18\cdot 10^{-6}$	$0.2 \cdot 10^{-7}$
10.028 414 61	$0.936\cdot 10^{-5}$	$0.94\cdot 10^{-4}$	$0.14\cdot 10^{-4}$
20.846 143 11	$0.532\cdot 10^{-2}$	0.11	$0.72\cdot 10^{-3}$
24.885 762 63	$0.800\cdot 10^{-3}$	$0.2\cdot 10^{-1}$	$0.31.10^{-4}$

Table 1

Computation with N = 8 gives approximations accurate to at least 10 digits (computed bound!) for all of the four eigenvalues.

The results for the other three types of symmetry are quite similar.

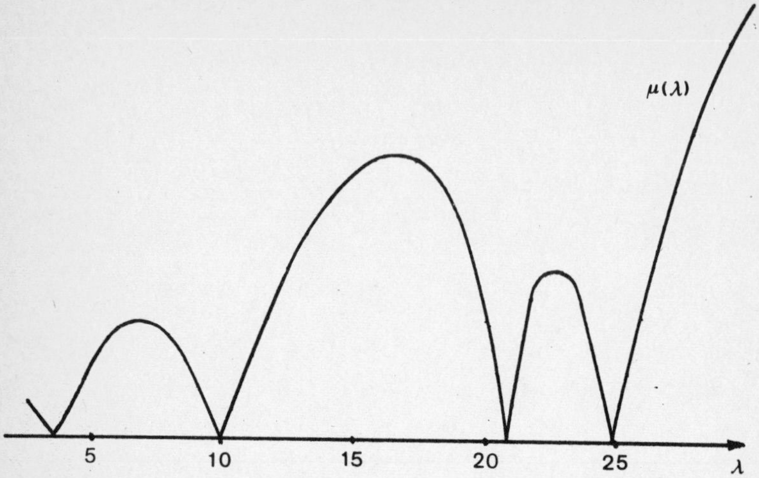

Figure 1

Example 2. Next we consider an L-shaped membrane as shown in Figure 2.

This problem is considerably more difficult due to the reentrant corner. We choose our approximating functions such that all of them are zero along the positive x_1- and the negative x_2-axis. This forces the approximating functions to have similar singularities in their derivatives as the eigenfunctions have. A simple analysis gives

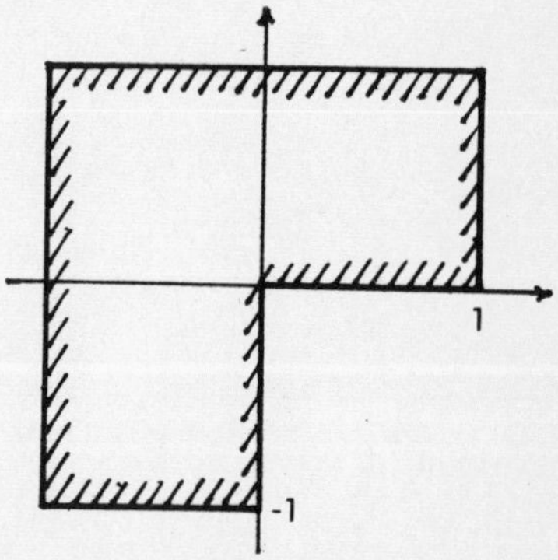

Figure 2

$$\phi_N(\lambda,z) = \sum_{l=1}^{N} \alpha_l J_{\frac{2}{3}l}(\sqrt{\lambda}r)\,\sin(\tfrac{2}{3}\ell\theta).$$

Again according to the symmetry-axis we could reduce the number of functions in advance and consider the two symmetry-types separately. We have not done that. Yet, by specifying the normalization condition to $\alpha_1=1$, we find eigenvalues with symmetric eigenfunctions (cf. Figure 3), requiring $\alpha_2=1$, we get the antisymmetric ones.

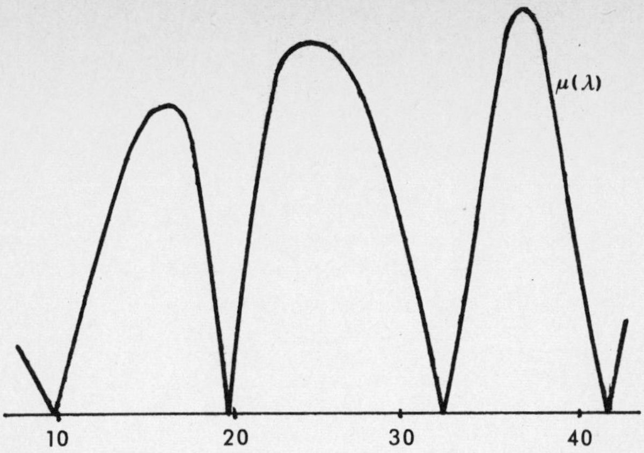

Figure 3

Example 3. As a last example we take again an L-shaped membrane but now without a symmetry axis, cf. Figure 4. Figure 5 gives the corresponding $\mu(\lambda)$.

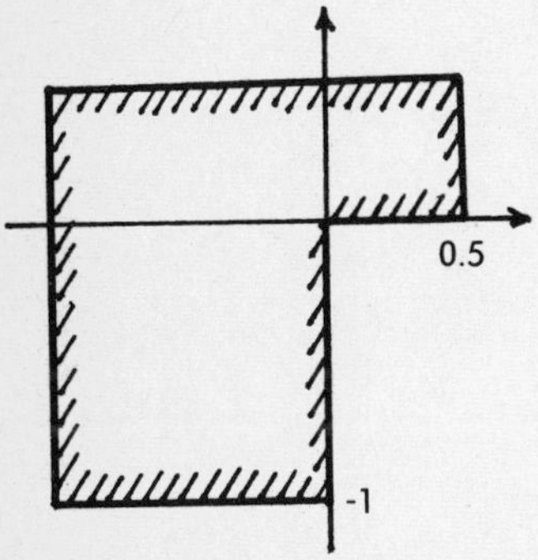

Figure 4

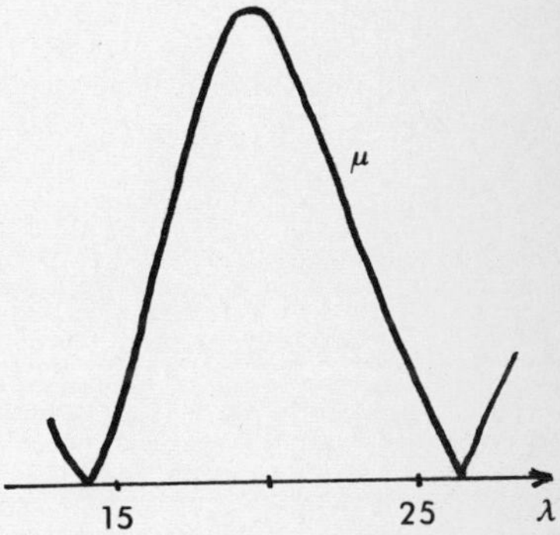

Figure 5

2. Generalized rational Chebyshev approximation.

2.A. Statement of problem.

The second problem we propose to treat by parametric linear programming is that of Chebyshev approximation by rational functions.

In its classical form, investigated already by Chebyshev, it is stated as follows:

For fixed $n,m\in\mathbb{N}$ let $a(p,q;x) = \sum_{i=1}^{n} p_i x^{i-1} / \sum_{i=1}^{m} q_i x^{i-1}$.

Then, given $f\in C[\alpha,\beta]$, $[\alpha,\beta]$ a compact interval, $\overline{p}_i,\overline{q}_i$ are to be determined such that the maximum-norm of the error function

$$\| f-a(p,q;\cdot)\|_\infty = \max_{\alpha\le x\le\beta} |f(x)-a(p,q;x)|$$

has minimal value subject to the condition that the denominator is non-vanishing, i.e. without loss of generality

$$\sum_{i=1}^{m} q_i x^{i-1}>0,\ x\in[\alpha,\beta].$$

We just collect some of the main theoretical results in Theorem 3 below. To formulate them, we need some notations:

Let $\overline{p},\overline{q}$ be a solution of the problem. We call $\overline{p},\overline{q}$ (or $a(\overline{p},\overline{q};\cdot)$) a <u>normal</u> solution if the polynomials $\sum_{i=1}^{n} \overline{p}_i x^{i-1}$ and $\sum_{i=1}^{m} \overline{q}_i x^{i-1}$ have no common factor and at least one of them has maximal degree n-1 or m-1 resp.

We call a point $(\overline{p},\overline{q})$ <u>critical</u>, if the zero function is a solution of the linearized approximation problem of approximating $f-a(\overline{p},\overline{q};\cdot)$ by elements in the linear space $T(\overline{p},\overline{q})\subset C[\alpha,\beta]$, tangent to the space of all rationals at $a(p,q;\cdot) = a(\overline{p},\overline{q};\cdot)$.

<u>Theorem 3.</u> A solution of the ordinary rational approximation problem always exists and is uniquely determined (as a function; of course the optimal $(\overline{p},\overline{q})$ is not!). $(\overline{p},\overline{q})$ is optimal if and only if it is critical. If $(\overline{p},\overline{q})$ is normal and we add some normalization condition like $\sum_{i=1}^{m} q_i \alpha^{i-1} = 1$, then $(\overline{p},\overline{q})$ is even locally strongly unique.

We mention that local strong uniqueness of the solution forms the basic tool in proving local superlinear convergence of a number of common methods like the nonlinear Remes-method, the linearization method, and the differential-correction method. In this paper we shall consider the generalized rational approximation problem, which rarely meets the condition of local strong uniqueness (cf. [12] for a more extensive discussion).

Let $B\subset\mathbb{R}^k$ be a compact set, $f;\varphi_i, i=1,\dots,n;\Psi_j, j=1,\dots,m$ given functions in C(B). Then instead of ordinary rationals on the interval as above we consider

$$a(p,q;x) = \sum_{i=1}^{n} p_i\varphi_i(x) \Big/ \sum_{j=1}^{m} q_j\Psi_j(x) \tag{2.1}$$

on the set B. For shorter notation we write

$$v(p;x) = \sum_{i=1}^{n} p_i\varphi_i(x),\ w(q;x) = \sum_{j=1}^{m} q_j\Psi_j(x). \tag{2.2}$$

Again, in first instance, we seek for $\overline{p},\overline{q}$ which minimize

$$\| f - a(p,q;\cdot)\|_\infty = \max_{x\in B} | f(x) - a(p,q;x)|$$

under the additional condition that $w(q,x)>0$, $x\in B$. For this problem not even the existence of solutions is guaranteed, as simple examples show. Therefore, we replace $w(q,x)>0$ in B by the stronger regularizing condition that for some constant $\gamma > 0$

$$1\le w(q,x)\le\gamma \quad \text{for all } x\in B. \tag{2.3}$$

By performing an obvious transformation, we may state the problem as the nonlinear (in general semi-infinite) programming problem:

RAP (Rational Approximation Problem):

Minimize $\ell(p,q,\mu) = \mu$ subject to

$$\left.\begin{aligned} f(x)\, w(q,x) - \mu w(q,x) - v(p,x)&\le 0 \quad (2.4a)\\ -f(x)\, w(q,x) - \mu w(q,x) + v(p,x)&\le 0 \quad (2.4b)\\ - w(q,x)&\le -1 \quad (2.4c)\\ w(q,x)&\le \gamma \quad (2.4d) \end{aligned}\right\}\ x\in B.$$

For problem RAP the existence of a solution is proved by standard arguments if there are paramters q fullfilling (2.4c,d) at all. Moreover, we will assume in the sequel, that there exists q^* such that (2.4c,d) hold with strict inequality for all $x\in B$, i.e.:

$$1< w(q^*,x)< \ ,\ x\in B. \tag{2.5}$$

Contrary to the case of ordinary rationals we can expect neither a unique solution $a(\overline{p},\overline{q};\cdot)$ nor an optimal parameter $(\overline{p},\overline{q})$ which is strongly unique after normalization.

However, it remains true, that $(\overline{p},\overline{q})$ is optimal if and only if it is critical. The linearization of problem RAP at a point $(p,q,\mu) = (p^o,q^o,\mu^o)$ is easily shown to be equivalent to the following one, which only depends on q^o,μ^o.

$\underline{RAP_\ell(q^o,\mu^o)}$. Minimize $l(p,q;\mu) = \mu$ subject to

$$\begin{aligned} f(x)\, w(q,x)-\mu_o w(q,x)-v(p,x)-\mu w(q_o,x)&\le 0\\ -f(x)\, w(q,x)-\mu_o w(q,x)+v(p,x)-\mu w(q_o,x)&\le 0\\ -\, w(q,x)&\le -1\\ w(q,x)&\le \gamma \end{aligned}$$

Thus we have (cf. [7]).

<u>Theorem 4.</u> $(\overline{p},\overline{q},\overline{\mu})$ solves RAP if and only if $(\overline{p},\overline{q},\mu=0)$ solves $RAP_\ell(\overline{q},\overline{\mu})$.

2.B. A parametric programming approach to rational Chebyshev approximation

Theorem 4 shows that, in principle, problem RAP might be solved by adapting parameters $(q^o,\mu^o)\in\mathbb{R}^{m+1}$ such that $RAP_\ell(q^o,\mu^o)$ has a solution $(p^o,q^o,\mu=0)$. This procedure of iteratively solving linearized problems $RAP_\ell(q^i,\mu^i)$ indeed leads to an algorithm quite similar to the original differential correction algorithm (cf.[2]). Despite of its generally excellent convergence in practice up to now superlinear convergence of the differential correction algorithm could be proved only in the strongly unique case ([1]). In the sequel we consider a simplification of $RAP_\ell(q^o,\mu^o)$ which depends only on a single parameter and forms the basis of an algorithm the superlinear convergence of which can be shown under much less restrictive assumptions.

Let us consider the following simplified linear problem, which is obtained from $RAP_\ell(q^o,\mu^o)$ by replacing $w(q_o,x)$ by 1 (and writing λ instead of μ^o):

<u>LRAP(λ)</u>. Minimize $l(p,q,\mu) = \mu$ subject to

$$f(x)\, w(q,x) - \lambda w(q,x) - v(p,x) - \mu\le 0 \qquad (2.6a)$$

$$-f(x)\, w(q,x) - \lambda w(q,x) + v(p,x) - \mu\le 0 \qquad (2.6b)$$

$$-\, w(q,x) \;\le -1 \qquad (2.6c)$$

$$w(q,x) \;\le\; \gamma \qquad (2.6d)$$

Then we have

Theorem 5. $(\bar{p},\bar{q},\mu=0)$ is a solution of $RAP_\ell(\bar{q},\lambda)$ if and only if it is a solution of $LRAP(\lambda)$, and, equivalently, if $(\bar{p},\bar{q},\lambda)$ solves RAP.

Proof. The set of feasible points of the form (p,q,o) (i.e. $\mu=0$) is the same for $RAP_\ell(\bar{q},\lambda)$ and $LRAP(\lambda)$. Next it is very easy to show that for points $(\bar{p},\bar{q},\mu=o)$ the Kuhn-Tucker condition is identical for both problems. Yet, as (2.5) implies that both problems meet Slater's condition, the Kuhn-Tucker condition is necessary and sufficient for a solution. Observing Theorem 4, this finishes the proof.

Let $\mu(\lambda)$ be the value of $LRAP(\lambda)$ and $\bar{\lambda}$ such that $\mu(\bar{\lambda})=0$. Suppose $(\bar{p},\bar{q},\mu=0)$ solves $LRAP(\bar{\lambda})$. Then Theorem 5 states that $(\bar{p},\bar{q})$ defines the rational function $a(\bar{p},\bar{q};x)$ of best approximation to $f(x)$ and $\bar{\lambda}$ is the maximal error of approximation:

$$\bar{\lambda} = \| f-a(\bar{p},\bar{q};\cdot)\|_\infty ,$$

i.e. the value of the problem RAP.

Theorem 6. Let $\lambda>0$ be given and $p(\lambda)$, $q(\lambda)$, $\mu(\lambda)$ be a solution of $LRAP(\lambda)$. Then, for the value $\bar{\lambda}$ of RAP we have

$$\min\{\lambda + \mu(\lambda), \lambda+\mu(\lambda)/\gamma\} \le \bar{\lambda} \le \| f-a(p(\lambda),q(\lambda);\cdot)\|_\infty .$$

This inclusion gives the possibility to control convergence during an iterative method for finding a zero of $\mu(\lambda)$ and supplies an excellent stopping rule in connection with an algorithm for finding the zero of $\mu(\lambda)$. As we will see, $\mu(\lambda)$ is such that we can apply a Newton- like method which usually converges superlinearly. Therefore, in spite of the rather laborious computation of the function values $\mu(\lambda)$, the resulting algorithms are very efficient as we will see lateron.

2.C. Example.

Again we will illustrate the method by means of an example. On the square $B = [0,1]\times[0,1]$ we approximate the function

$$f(x) = (2x_1^2 + 3x_2^2)e^{x_2^2 - x_1^2}$$

by rationals of type

$$a(p,q;x) = \frac{p_1+p_2x_2+p_3x_1+p_4x_1x_2}{q_1+q_2x_2+q_3x_2^2+q_4x_1+q_5x_1x_2+q_6x_1x_2^2+q_7x_1^2+q_8x_1^2x_2+q_9x_1^2x_2^2}$$

Figure 6 shows the function $\mu(\lambda)$. We see that in the zero $\bar{\lambda}$ which is the minimal value of the error-norm, μ is obviously not differentiable.

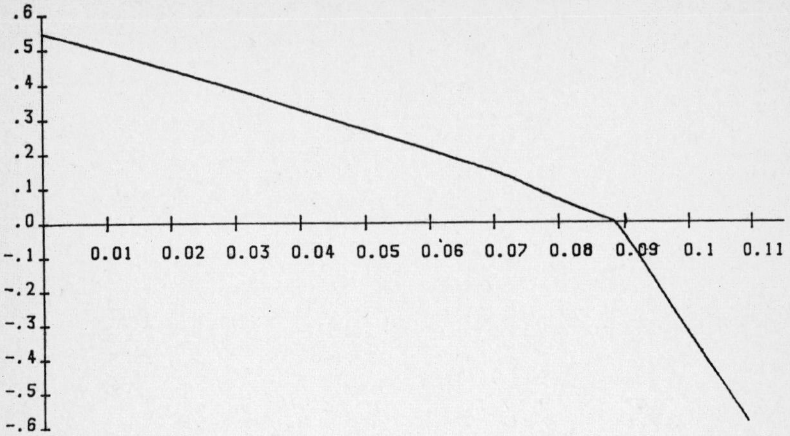

Figure 6

3.A general theorem on directional derivatives of $\mu(\lambda)$

In this section we consider a general one-parameter family of linear semi-infinite problems

$P(\lambda)$ Minimize $l(z) = c^T z$ subject to

$$z \in Z(\lambda) = \{z \in \mathbb{R}^n \mid z^T a(\lambda;x) \leq b(\lambda;x),\ x \in B\}$$

where λ varies in an open interval $I \subset \mathbb{R}$, $B \subset \mathbb{R}^m$ is a compact set, and for every fixed λ the functions $a(\lambda;\cdot):B \to \mathbb{R}^n$ and $b(\lambda;\cdot):B \to \mathbb{R}$ are continuous. We write $a(\lambda;\cdot) \in C(B,\mathbb{R}^n)$ and $b(\lambda,\cdot) \in C(B,\mathbb{R}) = C(B)$. Denoting by

$$v(\lambda) = \inf\{c^T z \mid z \in Z(\lambda)\}$$

the value of $P(\lambda)$, we will examine the question of one-sided differentiabilitiy of the function $v:I \to \mathbb{R}$.

Of course, to ensure differentiability of v we need corresponding assumptions on a,b as functions of λ (local conditions) and, moreover, some weak global assumptions ensuring that $Z(\lambda)$ depends in a stable way on λ.

Assumption 1: The mappings

$$\varphi_a: \begin{array}{l} I \to C(B,\mathbb{R}^n) \\ \lambda \to a(\lambda;\cdot) \end{array} \quad \text{and} \quad \varphi_b: \begin{array}{l} I \to C(B) \\ \lambda \to b(\lambda;\cdot) \end{array}$$

are continuously Frêchet-differentiable. (Instead we could require that the partial derivatives $a_\lambda(\lambda;x)$ and $b_\lambda(\lambda;x)$ of a and b are continuous functions of $(\lambda,x)\in I\times B$).

Assumption 2: In a given $\bar{\lambda}\in I$, the dual Problem $D(\bar{\lambda})$ is superconsistent, i.e.

$$-c\in \text{int } K[\{a(\bar{\lambda};x)\,|\,x\in B\}] \tag{3.1}$$

(where int K[M], $M\subset\mathbb{R}^n$, denotes the interior of the convex cone K[M] spanned by M), and $Z(\bar{\lambda})$ meets Slater's condition, i.e. there exists a $z^*\in\mathbb{R}^n$ such that

$$(z^*)^T\, a(\bar{\lambda};x)<b(\bar{\lambda};x),\ x\in B. \tag{3.2}$$

The results of the following theorem are standard (cf. [7]):

Theorem 7. Suppose that for $\bar{\lambda}\in I$ Assumptions 1 and 2 are fulfilled. Then the following assertions hold:

(i) There exists a neighborhood U of $\bar{\lambda}$, $U\subset I$, such that for $\lambda\in U$ problem $P(\lambda)$ has a nonempty, compact set $S^P(\lambda)\subset Z(\lambda)$ of solutions.

(ii) Let $\{\lambda_n\}\subset U$ be such that $\lambda_n\to\bar{\lambda}$ and z_n be such that $z_n\in S^P(\lambda_n)$ for all n. Then the sequence $\{z_n\}$ is bounded and its points of accumulation are in $S^P(\bar{\lambda})$. This, especially, implies that $v(\lambda)$ is continuous in $\bar{\lambda}$.

(iii) Let $M^+(B)$ be the set of nonnegative Borel-measures on B. Then for $\lambda\in U$ the dual problem

D(λ) Maximize $-\int_B b(\lambda;x)\,d\mu(x)$ subject to

$$\mu\in M^+(B) \quad \text{and} \quad \int_B a(\lambda;x)\,d\mu(x) = -c$$

has a nonempty set $S^D(\lambda)$ of solutions and we have strong duality between $P(\lambda)$ and $D(\lambda)$.

The next theorem is the main result of this section.

Theorem 8. Suppose that for $\bar{\lambda}\in I$ Assumptions 1 and 2 hold . Then $v(\lambda)$ possesses one-sided left- and right-hand-side derivatives $v'_-(\bar{\lambda})$ and $v'_+(\bar{\lambda})$ at the point $\bar{\lambda}\in I$. Moreover we have (with Fréchet-derivatives a_λ, b_λ)

$$v'_+(\bar\lambda) = \inf_{z\in S^P(\bar\lambda)} \sup_{\mu\in S^D(\bar\lambda)} \int_B (z^T a_\lambda(\bar\lambda;x) - b_\lambda(\bar\lambda;x))d\mu(x) \tag{3.3a}$$

and

$$v'_-(\bar\lambda) = \sup_{z\in S^P(\bar\lambda)} \inf_{\mu\in S^D(\bar\lambda)} \int_B (z^T a_\lambda(\bar\lambda;x)-b_\lambda(\bar\lambda;x))d\mu(x) \tag{3.3b}$$

In particular, this implies that $v(\lambda)$ is differentiable in $\bar\lambda$ if $P(\bar\lambda)$ and $D(\bar\lambda)$ have unique solutions.

Instead of presenting the rather technical proof (cf.[13]), we shall try to sketch intuitively two of the most common situations where $v(\lambda)$ fails to be differentiable.

We start our considerations at a point $\lambda_o \in U$ (with the U of Theorem 7) such that both the primal and dual problem $P(\lambda_o)$ and $D(\lambda_o)$ have a unique solution. Then it is well-known that the dual solution is uniquely represented by a measure $\mu_o \in M^+(B)$ with finite support $x_o=(x_o^1,\dots,x_o^k)$, $x_o^j \in B$, and strictly positive weights $\omega_o=(\omega_o^1,\dots,\omega_o^k)$, $\omega_o^j > o$. We write $\mu_o=(\omega_o,x_o)$.

Thus, let

$$S^P(\lambda_o) = \{z_o\} \quad , \quad S^D(\lambda_o) = \{(\omega_o,x_o)\}$$

Applying the rule of complementary slackness we get the equations

$$\varphi(\lambda_o;z_o;x_o^j) = z_o^T a(\lambda_o;x_o^j) - b(\lambda_o;x_o^j) = o, \quad j=1,\dots,k. \tag{3.4}$$

For simplicity, we assume that all points x_o^j are interior points of B. Then, from $z_o \in Z(\lambda_o)$ and (3.4) we conclude that the x_o^j are local maxima of $\varphi(\lambda_o;z_o;\cdot)$ (cf. Fig. 7a)). Therefore

$$z_o^T a_x(\lambda_o;x_o^j) - b_x(\lambda_o;x_o^j) = o, \; j = 1,\dots,k \tag{3.5}$$

(we assume that the partial derivatives a_x, b_x exist and are continuous). In addition, as Slater's condition holds, the Kuhn-Tucker-condition implies that

$$\sum_{j=1}^{k} \omega_o^j \, a(\lambda_o;x_o^j) = - c . \tag{3.6}$$

Thus, for $\lambda=\lambda_o$, our assumptions imply that (3.4)-(3.6) form a system of n+2k equations for just as much unknowns z,ω,x the unique solution of which is (z_o,ω_o,x_o). We write this system in condensed form as

$$F(\lambda;z,\omega,x) = 0 . \tag{3.7}$$

To resolve equations (3.7) w.r.t. (z,ω,x), suppose that at the point $(\lambda_0;z_0,\omega_0,x_0)$ the assumptions of the Implicite-Function-Theorem are met. Then, in a neighborhood $U(\lambda_0)$ of λ_0, by (3.7) continuously differentiable functions $z(\lambda),\omega(\lambda),x(\lambda)$ are defined. As long as both $z(\lambda)$ and $\mu(\lambda)=(\omega(\lambda), x(\lambda))$ remain feasible for $P(\lambda)$ and $D(\lambda)$ such that $\omega(\lambda)>0$ and $\{x^j(\lambda)\,|\,1\le j\le k\}$ includes all points $x\in B$ with $\varphi(\lambda;z(\lambda);\cdot) = 0$, $z(\lambda)$ and $\mu(\lambda)$ are the unique solutions of $P(\lambda)$ and $D(\lambda)$ resp. We assume that this is true for all λ in a maximal interval $I_0\subset U(\lambda_0)$.

Then, if λ_0 is an interior point of I_0, $v(\lambda)=c^Tz(\lambda)$ is obviously differentiable in λ_0 (actually, in accordance with Theorem 8, this is true for every interior point of I_0) and formula

$$v'_+(\lambda_0) = v'_-(\lambda_0) = \sum_{j=1}^{k} \omega_0^j[z_0^T a_\lambda(\lambda_0;x_0^j)-b_\lambda(\lambda_0;x_0^j)] \tag{3.8}$$

is an immediate consequence of the Implicite-Function-Theorem.

Let $\bar\lambda$ be the right endpoint of the interval I_0. We assume that $\bar\lambda$ is an interior point of $U(\lambda_0)$, i.e. $z(\lambda),\mu(\lambda)$ are defined also to the right of $\bar\lambda$ (cf. Figure 8). We discuss two typical situations implying nondifferentiability of $v(\lambda)$ at $\bar\lambda$.

Case A. For $\lambda>\bar\lambda$, $z(\lambda)$ is not a feasible point of $P(\lambda)$. From this we conclude that for $z=z(\bar\lambda)$ an additional point $\tilde x^{k+1}\in B$ becomes active, i.e. $\varphi(\bar\lambda;z(\bar\lambda),\tilde x^{k+1}) = 0$. Suppose that, going beyond $\bar\lambda$, one of the old points $x^j(\lambda)$, say $x^1(\lambda)$, is dropped from the active set for the optimal z while a new one near $\tilde x^{k+1}$ joins. Then a k-tupel $\tilde x^2(\lambda),\ldots,\tilde x^{k+1}(\lambda)$ forms the active set for the solution $\tilde z(\lambda)$ of $P(\lambda)$, $\bar\lambda<\lambda<\bar\lambda+\varepsilon$, for some $\varepsilon>0$. Thus, a situation like the one sketched in Figure 7 a)-c) is met. For $\tilde z(\lambda)$, $\tilde\omega^2(\lambda),\ldots,\tilde\omega^{k+1}(\lambda),\tilde x^2(\lambda),\ldots,\tilde x^{k+1}(\lambda)$ a system analoguous to (3.4)-(3.6) holds and an argumentation similar to the above one yields a differentiable representation $v(\lambda)=c^T\tilde z(\lambda)$ for $\lambda>\bar\lambda$. If $c^Tz_\lambda(\bar\lambda)$ does not happen to be equal to $c^T\tilde z_\lambda(\bar\lambda)$, $v(\lambda)$ is obviously not differentiable in $\bar\lambda$ (cf. Fig. 8a).

However, the left- and right-hand side derivatives exist and are given by

$$\begin{aligned} v'_-(\bar\lambda) &= c^Tz_\lambda(\bar\lambda) = \min\{c^Tz_\lambda(\bar\lambda),\ c^T\tilde z_\lambda(\bar\lambda)\} \\ v'_+(\bar\lambda) &= c^T\tilde z_\lambda(\bar\lambda)= \max\{c^Tz_\lambda(\bar\lambda),\ c^T\tilde z_\lambda(\bar\lambda)\}. \end{aligned} \tag{3.9}$$

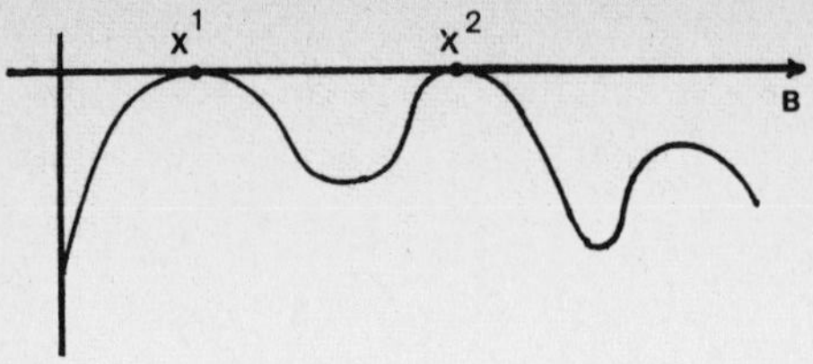

a) $\varphi(\lambda, z(\lambda);\cdot)$, $\lambda < \overline{\lambda}$

Case A

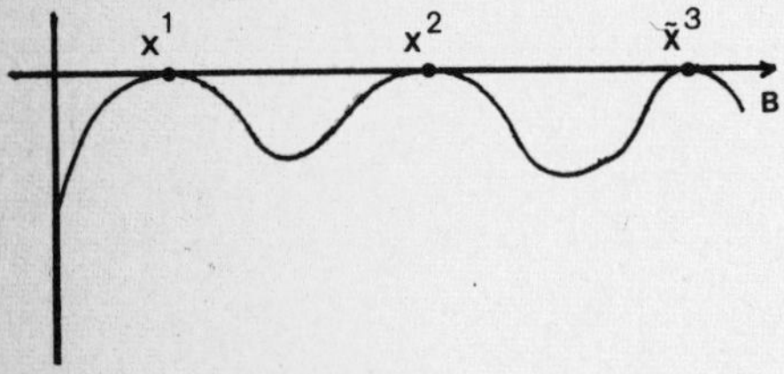

b) $\varphi(\lambda, z(\lambda);\cdot)$, $\lambda = \overline{\lambda}$

Case B

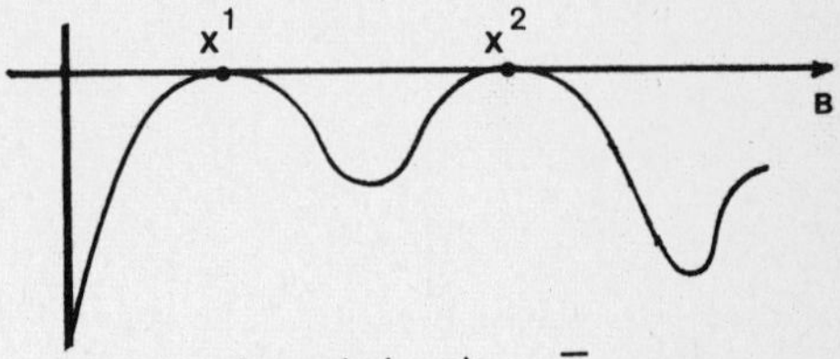

d) $\varphi(\lambda, z(\lambda),\cdot)$, $\lambda = \overline{\lambda}$

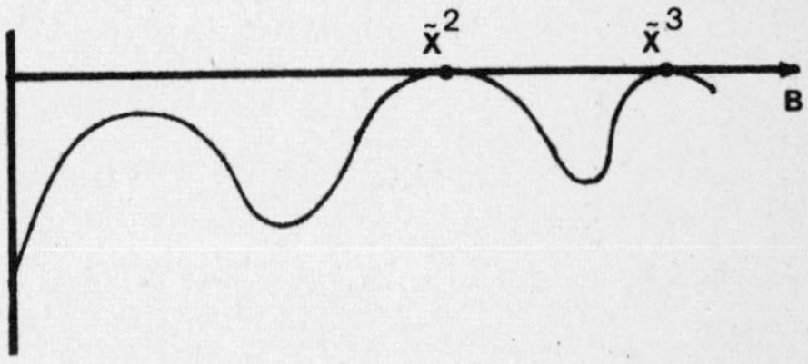

e) $\varphi(\lambda, \tilde{z}(\lambda),\cdot)$, $\lambda = \overline{\lambda}$

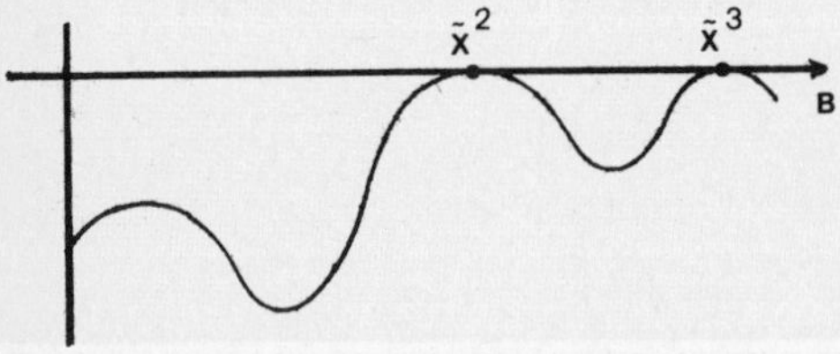

c) $\varphi(\lambda, z(\lambda);\cdot)$, $\lambda > \overline{\lambda}$

Figure 7

We emphasize that for $\lambda=\bar{\lambda}$ the dual problem $D(\bar{\lambda})$ has no longer a unique solution. Actually, every convex combination of the positive measures $\mu=(\omega(\bar{\lambda}), x(\bar{\lambda}))$ and $\tilde{\mu}=(\tilde{\omega}(\bar{\lambda}),\tilde{x}(\bar{\lambda}))$ solves $D(\bar{\lambda})$. Assuming $S^P(\bar{\lambda})=\{z(\bar{\lambda})\}$ $=\{\tilde{z}(\bar{\lambda})\}$, the discontinuity of the derivative is made plausible by a jump in the dual solution when passing $\bar{\lambda}$.

Case B. Now suppose that for $\lambda>\bar{\lambda}$ one of the Lagrangean parameters $\omega^j(\lambda)$, $\omega^1(\lambda)$ say, becomes negative. Then, by deleting the equations for x^1 from (3.4), (3.5) and the term with ω^1 from (3.6), we obtain a reduced system which, for $\lambda=\bar{\lambda}$, has solution $z(\bar{\lambda}),\omega^2(\bar{\lambda}),\ldots,\omega^k(\bar{\lambda}),\ldots,x^k(\bar{\lambda})$. If this system has a unique solution, then as a consequence of Theorem 8 we have differentiability of $v(\lambda)$ in $\lambda=\bar{\lambda}$ regardless of the fact that the number of active points changes when passing $\lambda=\bar{\lambda}$.

Therefore, suppose that the reduced system has a set of solutions, the projection onto the z-component of which is denoted by $S(\bar{\lambda})\subset\mathbb{R}^n$. Obviously, all feasible points of $S(\bar{\lambda})$ are solutions of $P(\bar{\lambda})$.

A typical situation is that $S(\bar{\lambda})\cap Z(\bar{\lambda})$ is an interval with extreme points (endpoints) $z(\bar{\lambda})$ and $\tilde{z}(\bar{\lambda})$ and that there is an additional active point $\tilde{x}^{k+1}(\bar{\lambda})$ for $z(\bar{\lambda})$, substituting $x^1(\bar{\lambda})$, such that $\tilde{x}^{k+1}(\lambda)$ remains active for $\lambda>\bar{\lambda}$. Thus we get a situation as sketched in Figure 7 d),e). Assuming $S^D(\bar{\lambda}) = \{\mu(\bar{\lambda})\}$, now the nondifferentiability of the value function (cf. Figure 8b)) is due to the jump of the primal solution in $\lambda=\bar{\lambda}$ (from one extreme point of $S(\bar{\lambda})\cap Z(\bar{\lambda})$ to another).

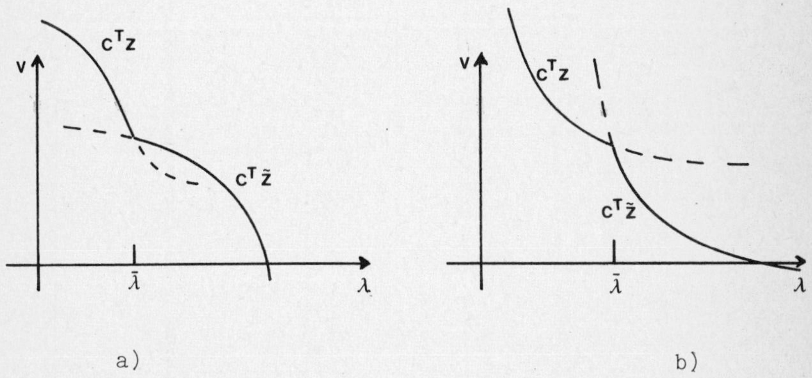

Figure 8

4. Application to rational approximation

It is obvious, that Assumption 1 is fulfilled for LRAP(λ). We show that under reasonable conditions on problem RAP problem LRAP(λ) meets Assumption 2, too.

Theorem 9. Suppose that each of the sets $\{\varphi_1,\ldots,\varphi_n\}$ and $\{\Psi_1,\ldots,\Psi_m\}$ of functions in C(B) defining the rational function is linearly independent and, moreover, that there exists a q* such that $1<w(q^*,x)<\gamma$, $x\in B$. Then Assumption 2 holds for LRAP(λ) and therefore $\mu(\lambda)$ has one-sided derivatives for all λ which are given by (3.3).

Proof. In [7,p71] it is shown that condition (3.1) of Assumption 2 is equivalent to the compactness of nonvoid levelsets $\{z|c^Tz\leq \text{const}, z\in Z(\lambda)\}$ of P(λ). Yet, it is very easy to show that all levelsets of LRAP(λ) are compact if $\{\varphi_i\}$ and $\{\Psi_j\}$ are linearly independent sets.

Taking q* as in the theorem, $p^*\in\mathbb{R}^n$ arbitrarily and $\mu^*> \| f(\cdot)w(q^*,\cdot)-v(p^*,\cdot)\|_\infty +|\lambda|\gamma$, then $z=(p^*,q^*,\mu^*)$ solves (3.2). Thus, Assumption 2 holds and Theorem 9 is proved.

For given $\bar\lambda$, let $\bar z = (\bar p,\bar q,\bar\mu)$, $(\bar\omega,\bar x)$ be solutions of P($\bar\lambda$) and D($\bar\lambda$). Then in (3.3) we obtain

$$\int_B [\bar z^T a_\lambda(\bar\lambda;x) - b_\lambda(\bar\lambda;x)]d\mu(x) = -\Sigma' \bar\omega^j w(\bar q;\bar x^j) \qquad (4.1)$$

where Σ' indicates summation over those j only for which $\bar x^j$ is active for (2.6a) or (2.6b).

Therefore, if P($\bar\lambda$), D($\bar\lambda$) have unique solutions, then $\mu(\lambda)$ is differentiable and $\mu'(\bar\lambda)$ is given by (4.1), an expression which requires no relevant additional computation.

This suggests to determine the zero of $\mu(\lambda)$ with some Newton-method

$$\lambda_{i+1} = \lambda_i+\mu(\lambda_i) \;/\; \Sigma' \omega^j_{(i)} w(q_{(i)};x^j_{(i)}) \qquad (4.2)$$

where $\omega^j_{(i)}$, $z_{(i)} = (p_{(i)},q_{(i)},\mu_{(i)})$, and $(\omega_{(i)},x_{(i)})$ are the computed solutions of P(λ_i) and D(λ_i) resp.

In case that $\mu(\lambda)$ was differentiable, we would have superlinear convergence. A simple consideration shows that in the zero-point $\bar\lambda$ of μ the function $\mu(\lambda)$ is in general not differentiable (cf.[12]). However, for the superlinear convergence of Newton's method it is sufficient that $\mu(\lambda)$ is continuously differentiable in some pointed neighborhood $U(\bar{\lambda})\setminus\{\bar{\lambda}\}$ of the zero-point $\bar{\lambda}$ and that, in addition, $\mu(\lambda)$ has one-sided derivatives from the left and the right at $\lambda=\bar{\lambda}$.

This is easily seen by considering the auxiliary function

$$\tilde{\mu}(\lambda) = \begin{cases} \mu(\lambda) & \text{if } \lambda\leq\bar{\lambda} \\ \mu(\lambda)\mu'_-(\bar{\lambda})/\mu'_+(\bar{\lambda}) & \text{if } \lambda>\bar{\lambda} \end{cases}$$

which is obviously continuously differentiable in a full neighborhood of $\bar{\lambda}$. Newton's method applied to $\tilde{\mu}$ is known to converge superlinearly and, on the other hand, yields the same iterates as if Newton's method was applied to μ.

For the resulting method global as well as linear convergence may be established just under the assumptions of Theorem 9. Moreover, under conditions much weaker than strong uniqueness of the solution, superlinear convergence can be shown (cf.[7,11,12]).

An algorithm based on the above has been implemented ([12]) in such a way that the basic inner linear programming cycle for solving problems LRAP(λ_i) has been fully integrated by using advanced stopping rules and restart methods within the outer loop of Newton-like iterations. According to (4.1) dual solutions of the respective linear problems are used to specify the Newton-steps. In addition the algorithm offers the opportunity of grid-refinement and grid-selection strategies (cf.[7]) to solve multivariate problems on very fine grids even if the available core is rather restricted.

To solve example 2.C. on a grid of 97 x 97 points the algorithm proceeds as follows:

≠ Points	≠ Points selected	≠ Newton-steps	≠ Simplexsteps
4 x 4	16	7	52
7 x 7	32	7	48
13 x 13	43	6	20
25 x 25	66	4	14
49 x 49	80	4	5
97 x 97	97	4	10

Thus, to solve this rational bivariate problem with 12 free parameters on 97 x 97 = 9.409 points required a total of 149 simplex-steps with on the average 220 constraints (4 for each point) in the linear-programming subproblems.

5. Application to the eigenvalue-problem

Again, for problem SIP(λ) in 1.C. and $\lambda > 0$ Assumption 1 holds. An argumentation similar to that of the last section shows that Assumption 2 holds if we select the functions from (1.9) in a way that their restrictions to B are linearly independent. Therefore, assuming the latter, again for all $\lambda > 0$ we have the existence of right- and left-hand derivatives of the value $\varepsilon(\lambda)$ of SIP(λ).

However, to get a similar result for $\mu(\lambda) = \varepsilon(\lambda)/\|\overline{\phi}_N(\lambda;\cdot)\|$ we must have differentiability of the optimal solution of SIP(λ) with respect to λ. I.e. the optimal parameters $\overline{\alpha}_i$, $\overline{\beta}_i$ in

$$\overline{\phi}_N(\lambda;z) = \sum_{i=1}^{N} [\overline{\alpha}_i \phi^c_{\nu_i}(\lambda;z) + \overline{\beta}_i \phi^s_{\nu_i}(\lambda;z)]$$

should be (one-sided) differentiable functions of λ_0 at the minimum $\overline{\lambda}$.

To be able to show that, we need stronger assumptions. If in the consideration following Theorem 8 the system (3.7) determines uniquely continuously differentiable functions $z(\lambda), \omega^j(\lambda)$, $x^j(\lambda)$, i.e. the assumptions of the Implicite-Function-Theorem are met in some $\lambda = \overline{\lambda}$, then we may conclude that the function $\mu(\lambda)$ we wish to minimize is differentiable in $\overline{\lambda}$. But in concrete cases we can say little in advance: we may not even be sure that SIP(λ) has a unique solution at $\lambda = \overline{\lambda}$.

Typically, around a local minimum $\mu(\lambda)$ looks like that given in Figure 9: We have used the following simple method to determine $\overline{\lambda}$:

Instead of $\mu(\lambda)$ we consider the function

$$\tilde{\mu}(\lambda) = \begin{cases} \mu(\lambda) & \text{if} \quad \lambda \le \overline{\lambda} \\ -\mu(\lambda) & \text{if} \quad \lambda > \overline{\lambda} \end{cases}$$

Given some λ, the decision if $\lambda < \overline{\lambda}$ or $\lambda > \overline{\lambda}$ is made by considering the sign of the difference-quotient with a neighboring point $\lambda+\varepsilon, \varepsilon$ very small.

Then, starting with an interval containing $\bar{\lambda}$, we simply apply regula falsi to determine the zero of $\tilde{\mu}(\lambda)$. In this way usually rather quickly an approximation is found with an error in the order of the jump $2\mu(\bar{\lambda})$ of $\tilde{\mu}$. For higher accuracy, the method is rather poor. Nevertheless, numerical examples usually lead to rather nice functions with a jump in the derivative at the minimum.

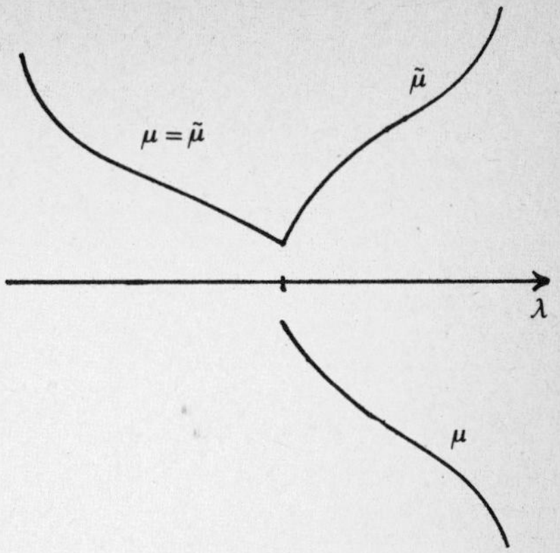

Figure 9

Figure 10 shows for the elliptic membrane that an exchange of active points takes place, when we pass the minimal point resp. the eigenvalue, implying nondifferentiability according to the discussion in Section 3. Figure 10 corresponds to the case N=5 in Example 1.

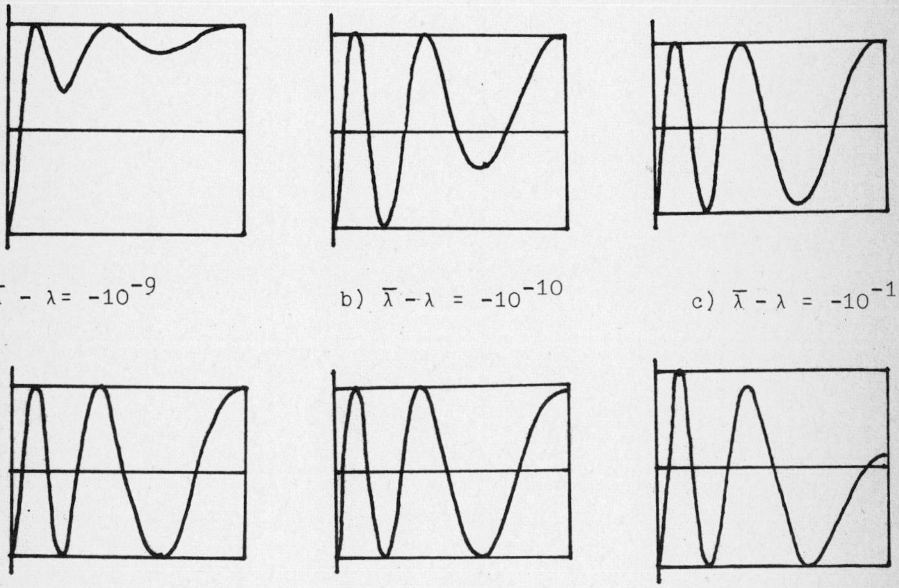

a) $\bar{\lambda} - \lambda = -10^{-9}$ b) $\bar{\lambda} - \lambda = -10^{-10}$ c) $\bar{\lambda} - \lambda = -10^{-11}$

d) $|\bar{\lambda} - \lambda| < 10^{-14}$ e) $\bar{\lambda} - \lambda = 10^{-11}$ f) $\bar{\lambda} - \lambda = 10^{-10}$

Figure 10: $\phi_5(\lambda,z)$ for $z \in B$ in Example 1

Examples: For the elliptic membrane of Example 1, one eigenvalue with eigenfunction symmetric to both axes is up to 12 figures given by $\lambda^* = 24.885\ 731\ 654\ 9$.

Taking $N = 5$ in ϕ_N^c of Example 1 the interval [24,25] is detected by a search with steplength 1. Application of the method above yields the table

λ	$\mu(\lambda)$	absol. crror
24.88683012	.8470914147E-03	.211E-01
24.88223052	.1022149930E-02	.254E-01
24.88474570	.8313994426E-03	.207E-01
24.88577817	.7993455501E-03	.199E-01
24.88527208	.8148410316E-03	.203E-01
24.88552755	.8068051392E-03	.201E-01
24.88565344	.8028454226E-03	.200E-01
24.88571594	.8008796105E-03	.199E-01
24.88574708	.7999000623E-03	.199E-01
24.88576263	.7994110588E-03	.199E-01

Taking now $N = 8$ and the interval [24.88,24.89] yields

24.88573150	.2502194672E-07	.623E-06
24.88573166	.3424075710E-09	.852E-08

References

[1] Barrodale, I.;Powell, M.J.D.; Roberts, F.D.K.: The differential correction algorithm for rational l_∞-approximation, SIAM J. Numer. Anal., 9 (1972), 493-504

[2] Cheney, E.W.; Loeb, H.L.: Two new algorithms for rational approximation, Numer. Math., 3 (1961), 72-75

[3] Courant, R.; Hilbert, D.: Methoden der mathematischen Physik, Band I, Springer, Berlin-Heidelberg-New York, 1968

[4] Donnelly, J.D.P.: Eigenvalues of membranes with rentrant corners, SIAM J. Numer. Anal., 6 (1969), 47-61

[5] Fox, L.; Henrici, P.; Moler, C.: Approximations and bounds for eigenvalues of elliptic operators, SIAM J. Numer. Anal., 4 (1967), 89-102

[6] Hersch, J.: Erweiterte Symmetrieeigenschaften von Lösungen gewisser linearer Rand- und Eigenwertprobleme, J.Reine u. Angew. Math., 218 (1965), 143-158

[7] Hettich, R.; Zencke, P.: Numerische Methoden der Approximation und semi-infiniten Optimierung, Teubner,Stuttgart, 1981

[8] Moler, C.; Payne, L.F.: Bounds for eigenvalues and eigenvectors of symmetric operators, SIAM J. Numer. Anal., 5 (1968), 64-70

[9] Nickel, K.: Extension of a recent paper by Fox, Henrici and Moler on eigenvalues of elliptic operators, SIAM J. Numer. Anal. 4 (1967), 483-488

[10] Piel, G.: Approximationsmethoden zur näherungsweisen Bestimmung von Eigenwerten und Eigenfunktionen elliptischer Differentialoperatoren, Diplomarbeit, Universität Bonn, 1981

[11] Speich, G.: Ein Algorithmus zur Lösung allgemeiner rationaler Approximationsprobleme. Thesis, University of Bonn, 1981

[12] Zencke, P.: A Newton-Differential-Correction-Algorithm for generalized rational Chebyshev approximation, in preparation

[13] Zencke, P.; Hettich, R.: A theorem on directional derivatives of the value of parametric semi-infinite programming problems, in preparation

DISTURBANCE ATTENUATION BY OUTPUT FEEDBACK

H.W.Knobloch
Mathematisches Institut
der Universität Würzburg
Am Hubland

D-8700 Würzburg

1. Introduction.

We consider in this paper linear time invariant control systems with a scalar input u, scalar output y and a scalar external disturbance signal v. The disturbance exerts its influence on the output via the plant. So we assume that the system description is given as follows

$$\dot{x} = Ax + Bu + Gv, \quad y = Cx. \tag{1.1}$$

x is the state variable; its dimension will be denoted by n.

Furthermore we will assume that v has a known dynamics. To be more precise: We allow specialization of v to an exponential polynomial of the form

$$\sum_{\nu} e^{\lambda_\nu t} p_\nu(t) \quad , \tag{1.2}$$

where the exponents λ_ν belong to a given finite set of complex numbers and the $p_\nu(\cdot)$ are polynomials of degree $\leq n_\nu$.

The complex numbers λ_ν, the non-negative integers n_ν and the matrices A,B,C,G constitute the information which is available for the solution of the problem to be discussed in the sequel. In precise terms it can be phrased as follows.

Find a strategy for specializing the control u which uses past observations of y only (i.e. the control $u(t)$ at a given time t depends upon the values of $y(s)$ for $t_0 \leq s \leq t$) and which achieves two goals.

(i) The output y decreases to zero as $t \to \infty$ irrespective of the initial state x_0 of the plant and any particular choice of the polynomials $p_\nu(\cdot)$ (cf. (1.2)). (1.3)

(ii) If $v = 0$ then the state x itself decreases to zero as $t \to \infty$ irrespective of the initial state x_0. In other words: The unperturbed system (1.1) is stabilized.

There is a classical way of dealing with this problem. One exploits the fact that the admissible specializations of the input variable v can be interpreted as output of a linear system, the so called disturbance model. Hence by state augmentation one can formally eliminate v from the right hand side of the dynamic equation and pass to a system description of the form

$$\dot{\tilde{x}} = \tilde{A}\tilde{x} + \tilde{B}u \ , \ y = \tilde{C}\tilde{x} . \qquad (1.4)$$

The uncertainty about the coefficients appearing in the disturbance signal (1.2) is then shifted to the uncertainty about the initial state $\tilde{x}(0)$.

The augmented system (1.4) is not stabilizable anymore, but it may still be possible - by applying a suitable control policy - to bring the output y asymptotically down to zero irrespective of the initial state. We then speak of output stabilization and note that it is one of the objectives which desired strategies should have (cf.(1.3)).

It is well known that output stabilization - if it is possible at all - can always be achieved by means of a feedback - observer structure. That is, one first stabilizes the output applying state-feedback $u = -F\tilde{x}$ and then replaces the inaccessible state $\tilde{x}$ by the state of a dynamical observer for the system (1.4). The same pattern of approach has been used for the construction of strategies which satisfy both conditions (1.3) (see e.g. [1] or [2],Ch.7). Therefore the problem we are concerned with here has in principle been settled within the framework of elementary control theory. From the viewpoint of applications however the answers which can be found in the existing literature are not satisfactory in various respects. Since it is required to design a classical feedback-observer-configuration for the augmented system - including a complete state-space model which generates the external signal v - the resulting closed loop structure becomes rather complex as soon as the disturbance signal is made more rich (and therefore more realistic) by adding exponential terms.

A new approach to the problem of disturbance attenuation has been proposed quite recently [3] and will be discussed further in this paper. Its advantage compared with the standard methods is computational simplicity which is not affected by the choice of "rich" disturbance models. The improvements which we are aiming at in the present contribution is twofold. (i) The construction of a certain convolution kernel - which plays the crucial role in the feedback strategy described in [3] - is simplified and involves - if carried out in the Laplace domain - rational operations only. (ii) We dispense with a series of restrictive assumptions underlying the analysis which was carried out in [3]. In fact our main result will hold under these very natural hypotheses concerning the dynamics of the plant and of the disturbance:

(i) The unperturbed system

$$\dot{x} = Ax + Bu, \quad y = Cx \tag{1.5}$$

is stabilizable and detectable.

(ii) The exponents λ_ν appearing in (1.2) satisfy the subsequent conditions.

$$\lambda_\nu \text{ is not a transmission zero of (1.5).} \tag{1.6}$$

$$\lambda_\nu \neq \lambda_\mu \text{ if } \nu \neq \mu, \ \mathrm{Re}(\lambda_\nu) \geq 0.$$

By a transmission zero we mean a zero of the formal numerator of the transfer function $u \to y$, that is the polynomial

$$\det(sI-A) \cdot C(sI-A)^{-1}B . \tag{1.7}$$

The paper is organized as follows. In the next section we state and comment our main result after the necessary preparations. The remaining sections are devoted to the proof which is given in various steps.

2. The main result.

We choose a row vector F_o and a column vector K_o such that all eigenvalues of the two matrices $A-BF_o$, $A - K_oC$ have negative real parts. This is possible, according to our hypothesis (1.5). We introduce the following polynomials

$$\chi_o(s) := \det(sI-(A-BF_o)), \ \chi_1(s) := \det(sI-(A-K_oC)),$$
$$k(s) := \chi_o(s)C(sI-(A-BF_o))^{-1}B . \tag{2.1}$$

<u>Lemma 2.1</u> If $k(\gamma) = 0$ and $\mathrm{Re}(\gamma) \geq 0$, then γ is a transmission zero of the system (1.5).

<u>Proof.</u> Let $H(s)$, $H_o(s)$ respectively be the transfer functions $u \to y$ of the system (1.5) and the system

$$\dot{x} = (A-BF_o)x + Bu, \quad y = Cx .$$

resespectively. We make use of the identity

$$H(s) = (1+F_o(sI-A)^{-1}B)H_o(s) .$$

Multiplying both sides with $\det(sI-A)$ one sees that the polynomial (1.7) is the product of $H_o(s)$ and a polynomial in s. Hence every zero of $H_o(\cdot)$ is a transmission zero. On the other hand $k(s) = \chi_o(s)H_o(s)$ and χ_o does not vanish in the closed right half plane. So any zero γ of k which has non-negative real part is in fact a zero of H_o. □

In the sequel polynomials p will play a role which depend upon several variables, one of them will always be s. By $\deg p$ we mean the degree with respect to s and we call p normalized if the leading coeffizient with respect to s is equal to 1.

Given a positive integer N we denote by

$$\underline{\zeta} = (\zeta_1, \ldots, \zeta_N) \quad \text{and} \quad \underline{\alpha} = (\alpha_1, \ldots, \alpha_{N+n})$$

respectively a N-tupel and a (N+n)-tupel respectively of different indeterminates. We put

$$\Delta(s,\underline{\zeta}) = \prod_{j=1}^{N} (s-\zeta_j), \quad \eta(s,\underline{\alpha}) = \prod_{i=1}^{N+n} (s-\alpha_i) . \tag{2.2}$$

<u>Lemma 2.2.</u> The polynomial equation

$$\eta(s,\underline{\alpha}) = k(s)p + \Delta(s,\underline{\zeta})q \tag{2.3}$$

has a unique solution

$$p = p(s,\underline{\alpha},\underline{\zeta}), \; q = q(s,\underline{\alpha},\underline{\zeta})$$

which satisfies the condition

$$\deg p < N, \quad q \text{ normalized}, \; \deg q = n. \tag{2.4}$$

p and q are polynomials in $s,\underline{\alpha}$ and rational functions in $s,\underline{\alpha},\underline{\zeta}$. To be more specific, p and q can be written as quotients of a polynomial in $s,\underline{\alpha},\underline{\zeta}$ and the resultant of Δ and k, that is the polynomial

$$R(\underline{\zeta}) := \prod_{j,\rho} (\zeta_j-\gamma_\rho). \tag{2.5}$$

(except for a constant factor). The product has to be extended over all ζ_j and all zeros γ_ρ of k with multiple zeros occurring repeatedly. Hence $R(\underline{\zeta}) = 0$ if and only if $k(\zeta_j) = 0$ for some component of $\underline{\zeta}$.

The proof is straightforward using the definition of the resultant of two polynomials (cf. e.g. [4], V, § 10). □

We now return to the general topic of this paper and consider a control system of the form (1.1). We also choose a class of disturbance signals of the form (1.2) by specifying the exponents λ_ν and the degrees n_ν of the polynomials $p_\nu(\cdot)$.

Theorem. Assume that the integer N and the variables ζ_j, α_i have been specialized subject to the following conditions.

$\mathrm{Re}(\alpha_i) < 0$, $i = 1,\dots, N+n$, $\mathrm{Re}(\zeta_j) \geq 0$, $j = 1,\dots,N$.
None of the ζ_j is a zero of k.
Every λ_ν appearing in (1.2) coincides with least 2^{n_ν} of the ζ_j. (2.6)

Let then the polynomials p and q be determined according to the conditions (2.3), (2.4) and let f(s) be the rational function

$$-\frac{\chi_o(s)p(s)}{q(s)\Delta(s)} \tag{2.7}$$

(the ζ_j are regarded now as fixed numbers, hence we have written $\Delta(s)$ instead of $\Delta(s,\underline{\zeta})$).

Claim. A feedback strategy which satisfies all requirements (1.3) can be set up in terms of the dynamic relations

$$u = -F_o x' + \int_0^t F(t-\tau)y(\tau)d\tau \ , \ \dot{x}'=Ax'+Bu+K_o(Cx'-y). \tag{2.8}$$

Here F(t) is the inverse Laplace-transform of the rational function (2.7).

Remarks. (i) One observes that (2.8) differs from the standard form of the dynamic equations for a regulator by a convolution integral. One can therefore interpret the conclusion of the theorem in the following way. The classical control law is simply supplemented by the convolution of the output with the kernel F(·) . Thereby the existing stabilizing property of the regulator is preserved and in addition the influence of the disturbance signal on the output is asymptotically removed.

(ii) The first hypothesis (2.6) is satisfied - in view of the second and of Lemma 2.1 - if none of the ζ_j is a transmission zero. It follows therefore from (1.6) that one can always specialize $\underline{\alpha}$ and $\underline{\zeta}$ in such a way that all requirements (2.6) are met.

(iii) We have $R(\underline{\zeta}) \neq 0$ because of (2.6). Hence the polynomial equation (2.3) is solvable subject to the side conditions (2.4). Note that these conditions imply that $f(s)$ is a proper rational function (cf.(2.7)), hence its inverse Laplace-transform is well defined.

(iv) The kernel function $F(\cdot)$ depends both upon the dynamics of the plant and that of the disturbance. This is clearly reflected in the factorization (2.7) of its Laplace-transform f. Note that χ_o depends only upon the plant, Δ only upon the disturbance whereas p and q are related both with plant and disturbance through the fundamental equation (2.3). Note also that p,q do not depend upon the complete state space model of the plant but only upon the polynomial k. In other words: The important data for the non-classical part of the control law are the disturbance modes and the transmission zeros of the plant.

(v) The control law depends upon $N+n$ parameters α_i which can be chosen freely except for the condition $Re(\alpha_i)<0$. These parameters determine the rate of the exponential decrease of the output and also - in case $v = 0$ - of the state, as can be seen from the following corollary. For this statement, by the way, it is not required that the eigenvalues of the matrices $A-BF_o$, $A-K_oC$ and the α_i have negative real parts.

Corollary. Consider the closed loop equation defined in terms of the two relations (1.1) and (2.8) and assume that the hypotheses of the theorem hold true. Then the output y can be written as a linear combination of exponentials $e^{\alpha' t}$ with polynomial coefficients, α' being a either a zero of η or an eigenvalue of one of the matrics $A - BF_o$ and $A - K_oC$. The coefficients depend upon the initial state of x,x' and of the disturbance signal (1.2). A representation of the same type exists for every component of x if the disturbance v is equal to zero.

3. Restatement of previous results.

As remarked before the two relations (1.1) and (2.8) constitute the equation of a closed loop system. It can also be written in the following way

$$\dot{x} = Ax - BF_o x' + Bu + Gv, \quad \dot{x}' = (A-BF_o)x' + Bu + K_o(Cx'-y), \quad y = Cx$$

$$u(t) = \int_0^t F(t-\tau)y(\tau)d\tau . \tag{3.1}$$

The two differential equations can be combined to give one equation of the form $\dot{\tilde{x}} = \tilde{A}\tilde{x} + \tilde{B}u + Gv$, $y = Cx$, where the eigenvalues of the matrix A are the eigenvalues of $A-BF_o$ and $A-K_oC$. So in order to prove our theorem it is sufficient to consider an equation of the form (1.1) together with a class of disturbance functions (1.2). We then have to construct a kernel function $F(\cdot)$ which is such that the application of the control law (3.1) leads to the effects described in the corollary, namely: The output y turns out to be a linear combination (with polynomial coefficients) of exponentials $e^{\alpha' t}$, where α' is either an eigenvalue of A or belongs to a prescribed set $\mathcal{A}$ of complex numbers. The same is true for the full state x in case the disturbance is zero.

An explicit construction of such a kernel has been carried out in [3] under the proviso that several hypotheses not mentioned so far are fulfilled. We repeat these conditions here:

(i) $v(t) = \sum_{\nu=1}^{N} a_\nu e^{\lambda_\nu t}$, where the a_ν are constants,

(ii) $\mathcal{A}$ consists of $N+n$ different complex numbers from the left half plane,

(iii) A is a stable matrix, the eigenvalues are simple and do not belong to the set $\mathcal{A}$,

(iv) the polynomial

$$\tilde{k}(s) := \chi(s)H_u(s), \text{where } \chi(s) := \det(sI-A), H_u(s) := C(sI-A)^{-1}B, \quad (3.2)$$

has no multiple zeros and does not vanish for $s=\lambda_\nu, \nu=1,\ldots,N$.

It was also assumed in [3] that the λ_ν are pure imaginary numbers, however this assumption was never used.
The definition of $F(\cdot)$ given in [3] then goes as follows.
$F(\cdot)$ is the inverse Laplace-transform of the rational function

$$f := \frac{1}{H_u}\left(1 - \frac{\tilde{\eta}}{H_u \chi \Delta q}\right) \quad (3.3)$$

where the various symbols have the following meaning:

H_u, χ : cf. (3.2),

$$\Delta(s) := \prod_{\nu=1}^{N} (s-\lambda_\nu), \quad \tilde{\eta}(s) := \prod_{\alpha\in\mathcal{A}} (s-\alpha). \quad (3.4)$$

Note that H_u is the transfer function $u \to y$ of the system

$$\dot{x} = Ax + Bu, \; y = Cx .$$

Finally q is a rational function with poles at the zeros of $\tilde{k}$ which satisfies the subsequent condition: $q-\tilde{\eta}/\Delta\tilde{k}$ vanishes at $s = \infty$ and has its poles at the zeros of Δ (cf. [3],(41),(45),(47)). $\tilde{k}$ is the polynomial introduced by (3.2). One sees now immediately that q can be determined from these properties. In fact, write q as $\tilde{q}/\tilde{k}$, then the relations

$$\tilde{\eta} = \tilde{k}\tilde{p} + \Delta\tilde{q}\ , \ \deg\tilde{p} < N = \deg \Delta \tag{3.5}$$

have to be satisfied with some further polynomial $\tilde{p}$. Since $\tilde{k}$ and Δ are relatively prime, there exists one and only one pair of polynomials $(\tilde{p},\tilde{q})$ for which condition (3.5) holds (cf. Section 2). Using (3.2), (3.5) one can simplify the expression on the right hand side of (3.3). So we can write the Laplace-transform of $F(\cdot)$ as

$$f = - \frac{\chi\tilde{p}}{\Delta\tilde{q}}\ . \tag{3.6}$$

There is an alternative interpretation of the control law (3.1) in terms of a conventional feedback-observer structure. Consider the augmented state-space model for the plant and the disturbance

$$\dot{x} = Ax + Bu + G\sum_{\nu=1}^{N} \xi_\nu, \dot{\xi}_\nu = \lambda_\nu\xi_\nu, \nu = 1,\dots,N,\ y = Cx\,. \tag{3.7}$$

One can then find a special dynamical observer for (3.7) (actually a collection of decoupled first-order reduced observers, cf. [3],(24)) with the subsequent properties. The state $\hat{\xi}$ of the observer is a $(N+n)$-dimensional vector which satisfies the differential equation

$$\dot{\hat{\xi}} = \hat{A}\hat{\xi} + \hat{B}u + \hat{D}y, \tag{3.8}$$

where $\hat{A}$ is the diagonal matrix made up of the elements of $\mathcal{A}$, $\hat{B}$ the column vector with elements $-H_u(\alpha), \alpha\in\mathcal{A}$, and $\hat{D}$ the vector $(1,1..)^T$. The observer property is reflected in the fact that

$$\hat{\xi} - Px - Q\xi\ , \quad \xi := (\xi_1,\dots,\xi_N)^T, \tag{3.9}$$

satisfies the homogeneous equation $\dot{\hat{\xi}} = \hat{A}\hat{\xi}$, where P and Q are suitable matrices. This is true for $y = Cx$ and irrespective of the specialization of the input function $u(\cdot)$. The observer is just constructed for the purpose of output stabilization. As has been shown in [3], Sec. 3, one can find a row vector F such that

$$\hat{F}P = 0,\ \hat{F}Q = (\vartheta(\lambda_1),\dots,\vartheta(\lambda_N)) \tag{3.10}$$

where $\vartheta(s)$ is a certain rational function (the quotient of the transfer functions $u \to y$, $v \to y$ of the system (1.1)). Furthermore the output of (3.7) is stabilized by applying the state feedback

$$u = -\sum_{\nu=1}^{N} \vartheta(\lambda_\nu)\xi_\nu . \tag{3.11}$$

In other words: If u is specified according to (3.11) then y decreases to zero for $t \to \infty$, irrespective of the initial state $x(0), \xi(0))$. Obviously x itself is stabilized if $\xi(0) = 0$, since A is a stable matrix. From what was said above in connection with (3.9) it is then clear that one arrives at the same end by applying the output feedback

$$u = -\hat{F}\hat{\xi}. \tag{3.12}$$

For later purposes we note that - for fixed $\underline{\alpha}$ - the components of $\hat{F}$ are rational functions of $\lambda_1, \ldots, \lambda_N$ (cf. [3],(28)-(32)).

That (3.12) is nothing else than (3.1) can now be demonstrated by some standard arguments as has been done in [3], Sec.4. In fact we have

Lemma 3.1. If $y(\cdot)$ is any continuous function and $\hat{\xi}(\cdot)$ solution of the initial value problem

$$\dot{\hat{\xi}} = \hat{A}\hat{\xi} - \hat{B}\hat{F}\hat{\xi} + \hat{D}y(t), \quad \hat{\xi}(0) = 0,$$

then $-\hat{F}\hat{\xi}(t) = \int_0^t F(t-\tau)y(\tau)d\tau$, where $F(\cdot)$ is the inverse Laplace-transform of f (cf. (3.6)).

Summarizing what we found so far we have

Proposition 1. Assume that the hypotheses (i)-(iv) mentioned at the beginning of this section are satisfied and let $H_u, \tilde{k}, \chi, \Delta, \tilde{n}, \tilde{p}, \tilde{q}, f$ be defined according to (3.2)-(3.6). Let $F(\cdot)$ be the inverse Laplace-transform of f. Then the output $y = Cx$ of the system

$$\dot{x} = Ax + Bu + Gv$$

with

$$u = \int_0^t F(t-\tau)y(\tau)d\tau, \quad v = \sum_{\nu=1}^{N} a_\nu e^{\lambda_\nu t} \tag{3.14}$$

decreases to zero as $t \to \infty$ irrespective of the values for the a_ν and the initial state $x(0)$. If all a_ν are zero then the same statement holds true for x itself.

For a proof see [3], Proposition 4 and the subsequent considerations.

One can go a step further and specify the exponentials which appear in the representation of y and x. To this purpose we remark that - according to Lemma 1 - x can be viewed upon as part of the state vector of the closed-loop system

$$\dot{x} = Ax - B\hat{F}\hat{\xi} + G\sum_{\nu=1}^{N}\xi_\nu, \quad \dot{\xi}_\nu = \lambda_\nu\xi_\nu, \nu=1,\dots,N, \quad \dot{\hat{\xi}} = \hat{A}\hat{\xi} - \hat{B}\hat{F}\hat{\xi} + \hat{D}Cx.$$

We introduce $\hat{\xi}-Px-Q\xi$ instead of $\hat{\xi}$ as state variable and make use of what has been said in connection with (3.9) and of (3.10). It is then easy to see that with respect to the new coordinates the system matrix can be written in this form

$$\begin{pmatrix} A & * & * \\ 0 & \Lambda & 0 \\ 0 & 0 & \hat{A} \end{pmatrix}$$

where $\Lambda = \text{diag}(\lambda_1,\dots,\lambda_N)$. Hence any eigenvalue of this matrix is simple and coincides with an eigenvalue of A or with one of the λ_ν or with an element of the set $\mathcal{A}$. So each component of x is a linear combination (with constant coefficients) of exponentials $e^{\lambda_\nu t}$ or $e^{\alpha t}$ where α is an eigenvalue of A or an element of the set $\mathcal{A}$. Now the λ_ν have all non-negative real parts, hence $e^{\lambda_\nu t}$ cannot appear in any linear combination which vanishes for $t\to\infty$.

Corollary. The conclusion of proposition 1 can be phrased in this way: The output y is a linear combination (with constant coefficients) of exponentials $e^{\alpha t}$, where α belongs to $\mathcal{A}$ or is an eigenvalue of A. The same is true for the components of x if all a_ν are zero.

4. Proof of the theorem

We begin with some general remarks concerning initial value problems of the form

$$\dot{x}(t) = Ax(t) + \int_0^t L(t-\tau)x(\tau)d\tau, \quad x(0) = x_o. \qquad (4.1.)$$

Here $L(\cdot)$ is some matrix which depends, say, continuously upon t and which has a Laplace-transform $\mathcal{l}$. (4.1) can be written as a Volterra integral equation

$$x(t) = x_o + \int_0^t (Ax(\tau) + z(\tau))d\tau, \quad z(t) = \int_0^t L(t-\tau)x(\tau)d\tau$$

and then solved using standard Laplace-transform technique. As a result one obtains a representation of the solution in the form

$$x(t) = M(t)x_0$$

where $M(\cdot)$ is the inverse Laplace-transform of a certain submatrix of

$$\frac{1}{s}\begin{pmatrix} I-\frac{1}{s}A & -\frac{1}{s}I \\ -l(s) & I \end{pmatrix}^{-1} . \tag{4.2}$$

We now return to the discussion of the system (3.7). If a control law of the form (3.1) is applied one arrives at a differential-integral equation of the form (4.1) where the vector (x,ξ) now plays the role of x and the matrix (4.2) assumes the form

$$\frac{1}{s}\left(\begin{array}{cc|c} I-\frac{1}{s}A & -\frac{1}{s}G,\dots,-\frac{1}{s}G & \\ 0 & I-\frac{1}{s}\Lambda & -\frac{1}{s}I \\ \hline 0 & -Bf(s)C & \\ 0 & 0 & I \end{array}\right)^{-1} =: \begin{array}{c} n\{ \\ \\ \end{array}\left(\begin{array}{c|c|} \overbrace{m_0(s)}^{n} & \overbrace{m_1(s)}^{N} \\ \hline & \\ & \end{array}\right). \tag{4.3}$$

The solution $x(\cdot)$ of (3.7) can then be expressed in terms of the initial values as

$$x(t) = M_0(t)x(0) + M_1(t)\,\xi(0)\ , \tag{4.4}$$

where $M_i(\cdot)$ is the inverse Laplace-transform of $m_i(\cdot)$, $i = 0,1$. $m_0(s)$ and $m_1(s)$ are those matrices of type (n,n) and (n,N) which can be obtained by partitioning (4.3) in the way indicated above. The statement of the corollary to Proposition 1 can now be phrased as follows

> The elements of $M_0(t)$ and of $CM_1(t)$ are linear combinations (with constant coefficients) of exponentials $e^{\alpha t}$ (4.5)
> where α is either an eigenvalue of A or an element of $\mathcal{A}$.

Since the eigenvalues of A are simple and different from the elements of $\mathcal{A}$ this property of $M_0(\cdot)$ and $CM_1(\cdot)$ is equivalent with the following property of $m_0(\cdot)$, $Cm_1(\cdot)$:

> The elements of these matrices are polynomial fractions with denominator $\chi_{\tilde{n}}$.

We are now in a position to carry out the essential step in the proof of our theorem. To this purpose we first note that the polynomials, rational functions and matrices introduced so far, namely

$$\chi, \Delta, \tilde{p}, \tilde{q}, f, m_o, m_1$$

are well defined under the only assumption that $\tilde{k}$ and Δ are relatively prime (cf. (3.2),(3.5),(3.6),(4.3)). The next proposition essentially states that nothing more than this hypothesis is required in order to make sure that (4.5) holds true (except for the fact that the coefficients need not be constants anymore but may become polynomials).

Proposition 2. If Δ and $\tilde{k}$ are relatively prime then the matrix and the row vector

$$\chi\tilde{\eta}\, m_o \quad \text{and} \quad \chi\tilde{\eta}Cm_1 \tag{4.6}$$

have as elements polynomials of degree $< N' := \deg(\tilde{\eta}\chi)$.

Proof. Consider the Laurent-expansion at $s = \infty$ of the matrix (4.3) after multiplication with $\chi\tilde{\eta}$. It is easy to see from the structure of the matrix (4.3) and the rational function f that this expansion assumes the form

$$\sum_{\mu < N'} s^{\mu} V_{\mu} \tag{4.7}$$

where the elements of the matrices V_{μ} are polynomials in the elements of the matrices A,B,C,Λ,G and the coefficients of the polynomials $\chi,\Delta,\tilde{p},\tilde{q}$. The latter ones are polynomials in the $\alpha \in \mathcal{A}$, the λ_{ν} and the coefficients of $\tilde{k}$ divided by the resultant $R(\tilde{k},\Delta)$ of $\tilde{k}$ and Δ (cf. Section 2). The coefficients of $\tilde{k}$ finally are polynomials in the elements of A,B,C (cf. (3.2)).

We denote the collection of these quantities by σ and regard σ as a parameter which can vary freely subject to the condition $R(\tilde{k},\Delta) \neq 0$. We then consider the Laurent-expansion of the matrices (4.6) at $s=\infty$:

$$\chi(s)\tilde{\eta}(s)m_o(s) = \sum_{\mu<N'} s^{\mu}V_{\mu}^{(0)}(\sigma) \ , \qquad \chi(s)\tilde{\eta}(s)Cm_1(s) = \sum_{\mu<N'} s^{\mu}V_{\mu}^{(1)}(\sigma)$$

where $V_{\mu}^{(i)}(\sigma)$ are suitable submatrices of V_{μ} and hence also rational functions of σ. If σ is specialized in such a way that all hypotheses (i)-(iv) listed at the beginning of Section 3 are satisfied then $V_{\mu}^{(i)}(\sigma) = 0$ whenever $\mu<0, i=0,1$: This is, as we have remarked above, the translation of the statement (4.5) into the Laplace domain. Since the restrictions imposed by the said hypotheses allow specialization of σ in some open set it is clear that $V_{\mu}^{(i)}(\sigma) = 0$ identically in σ whenever $\mu<0$. Thereby the proposition is proved. □

From the proposition one can easily infer that the statement of our theorem holds true in case the coefficients $p_\nu(\cdot)$ in the representation (1.2) of v are constants. In fact, if the disturbance signal v is of the form

$$\sum_\nu a_\nu e^{\lambda_\nu t} \tag{4.8}$$

and if the augmented system equation (3.1) (plant plus standard observer) is supplemented by a dynamical generator for (4.8) one arrives at an equation of the form (3.7). It is easy to realize that the polynomials χ and $\tilde{k}$ respectively turn out to be equal to $\chi_o\chi_1$ and $k\chi_1$ respectively (cf. (2.1),(3.1),(3.2)). We now choose as $\tilde{\eta}$ the product of χ_1 and a Hurwitz-polynomial η of degree $N+n$ (note that the dimension of the state variable for the augmented system is $2n$). From (3.5) one then sees immediately that $\tilde{q}$ is divisible by χ_1. If the common factor is cancelled on both sides of (3.5) then this equation - except for the notation - is nothing else than (2.3). One also can cancel the factor χ_1 in the numerator and denominator of the representation (3.6) and then the function f as defined by (3.6) turns out to be the same as the f introduced by (2.7).

We now consider a disturbance signal of the general form (1.2) (i.e. the coefficients of the exponentials are polynomials) and wish to extend the foregoing results to this situation. We use the symbol $\underline{\lambda}$ from now on in order to denote the N-tupel $(\lambda_1,\ldots,\lambda_N)^T$ and put

$$\Delta(s,\underline{\lambda}) := \sum_{\nu=1}^{N} (s-\lambda_\nu).$$

The system equation underlying our considerations is of the form (1.1) and the control law is again

$$u = \int_0^t F(t-\tau)y(\tau)d\tau$$

where $F = F(t,\underline{\lambda})$ is defined as in Section 3 (cf.(3.6)). Note that (3.6) makes sense under the only condition that each λ_ν is not a zero of $\tilde{k}$. $\tilde{\eta}$ is a fixed polynomial of degree $N+n$.

For notational convenience we will use in the sequel the symbol $\mathcal{H}$ in order to denote the set of all scalar functions φ of t and $\underline{\lambda} = (\lambda_1,\ldots,\lambda_N)^T$ which can be represented as finite linear combinations of products $e^{\alpha t}t^\rho\pi(\lambda)$, where α is a zero of $\tilde{\eta}\chi$, ρ a non-negative integer and $\pi(\cdot)$ a quotient of a polynomial in $\underline{\lambda}$ and some power of the resultant of Δ and $\tilde{k}$. $\underline{\lambda}$ for the time being has to be regarded as a N-tupel of independent indeterminates.

We now are in the position to state the conclusive result of this section from which the full statement of our theorem follows immediately.

<u>Proposition 3.</u> If the first 2^r components of $\underline{\lambda}$ are equal, i.e. if

$$\Delta(s,\underline{\lambda}) = (s-\lambda_1)^{2^r}\Delta_1(s,\underline{\lambda}) \tag{4.9}$$

then the solution $x(\cdot)$ of the initial value problem

$$\dot{x}(t) = Ax(t) + B\int_0^t F(t-\tau)Cx(\tau)d\tau + at^r e^{\lambda_1 t}, \quad x(0) = 0,$$

satisfies the condition $y(\cdot) = Cx(\cdot) \in \mathcal{H}$.

<u>Proof.</u> By induction with respect to r. If $r = 0$ then the statement can be inferred from Proposition 2 (or rather from the remark following the proof).
So let us now assume that the proposition has been proved for some $r-1 \geq 0$. The polynomial (4.9) is then obtained from the polynomial

$$\hat{\Delta}(s,\underline{\lambda}) := (s-\lambda_1)^{2^{r-1}}(s-\lambda_2)^{2^{r-1}}\Delta_1(s,\underline{\lambda})$$

by letting λ_2 be equal to λ_1. Let $x_i(\cdot)$ be the solution of the initial value problem

$$\dot{x}(t) = Ax(t) + B\int_0^t \hat{F}(t-\tau)Cx(\tau)d\tau + at^{r-1}e^{\lambda_i t}.$$

$i = 1,2$. $\hat{F}$ here is the kernel function associated with the polynomial $\hat{\Delta}$. By hypothesis of induction $Cx_i(\cdot) \in \mathcal{H}$ and the same is true for the partial derivatives with respect to λ_ν. It is now easy to see that the function

$$x(\cdot) = (\partial/\partial\lambda_1)x_1(\cdot) - (\partial/\partial\lambda_1)x_2(\cdot)$$

has the desired properties if $\lambda_1 = \lambda_2$. □

Reviewing this result and its implications for our theorem one gets the impression that it is too crude. What one would expect is that the multiplicity of λ_1 as a zero of Δ need not be greater than the degree of the polynomial $p_1(\cdot)$ which appears as coefficient of $e^{\lambda_1 t}$. So there remains an obvious desideratum, namely to determine the minimal multiplicity which each λ_ν must have as a zero of Δ in order that the strategy described in (2.8) achieves the objectives (1.3).

References

[1] B.A.Francis: The linear multivariable regulator problem. SIAM J.Control and Optimization 15 (1977), pp. 486-505.

[2] H.W.Knobloch and H.Kwakernaak: Lineare Kontrolltheorie. To appear. Springer Verlag, Berlin, Heidelberg, New York, Tokyo 1984/85.

[3] H.W.Knobloch: On the principle of "internal modelling" in linear control theory in Selected Topics in Operations Research and Mathematical Economics, G.Hammer and D.Pallaschke eds., Lecture Notes in Economics and Mathematical Systems Vol. 226, Springer Verlag Berlin, Heidelberg, New York, Tokyo 1984, pp. 131-151.

[4] S.Lang: Algebra, Addison-Wesley Publishing Company, Reading Ma, 1965.

TIME-REVERSAL OF DIFFUSION PROCESSES AND NON-LINEAR SMOOTHING

E. PARDOUX
UER de Mathématiques
Université de Provence
3, place Victor Hugo
13331 Marseille Cedex 3

Abstract: The object of this paper is to exploit recent results [4] on time-reversal of diffusion processes, in order to obtain new equations for the non-linear smoothing problem. We thus generalize some results that are known in the linear case.

1. Introduction :

In this paper, we consider the following non linear smoothing problem. Let $\{(X_t, Y_t), 0 \leq t \leq 1\}$ satisfy :

$$(1.1) \quad \begin{aligned} dX_t &= b(t, X_t)dt + \sigma(t, X_t)dW_t, \ X_0 \text{ given} \\ dY_t &= h(t, X_t)dt + g(t)dW_t + \tilde{g}(t)d\tilde{W}_t, \ Y_0 = 0 \end{aligned}$$

where $\{W_t\}$ and $\{\tilde{W}_t\}$ are independent Wiener processes. Precise assumptions will be stated below in section 2 and 3.

We want to characterize the conditional law of $X_t (t \in [0,1])$, given $\{Y_s, s \in [0,1]\}$. This problem has been considered recently in particular in PARDOUX [8], ANDERSON-RHODES [1] and PARDOUX [10]. In [8], we expressed the (unnormalized) conditional density of the smoothing problem in terms of the solution of the forward Zakaï equation of non linear filtering, and of the solution of an adjoint backward stochastic PDE, which is the Radon-Nikodym derivative of the (unnormalized) smoothing conditional law with respect to the (unnormalized) filtering conditional law.

Using the results of HAUSSMANN-PARDOUX [4], we shall formulate a backward smoothing problem - with "initial" conditions at final time $t = 1$ - equivalent to (1.1). This will lead us to introduce a backward Zakaï equation, and a forward adjoint SPDE, by the same arguments as in [9]. This gives us a second expression for the unnormalized conditional density of the smoothing problem. Using relations between the latter quantities and those associated with the initial formulation, in terms of the density of the a-priori law of X_t, we deduce two other expressions for the (unnormalized) conditional density of the smoothing problem.

Those proofs of some technical results which are not given below will be published elsewhere.

Notation :

Let $\{W_t, t \in [0,1]\}$ be a standard Wiener process defined on a probability space (Ω,F,P). For $t \in [0,1]$, define $G_t = \sigma\{W_s, 0 \leq s \leq t\}$ and $G^t = \sigma\{W_s - W_t, t \leq s \leq 1\}$. Here and in the sequel, we assume that all σ-algebras contain all P-null sets of the largest σ-algebra F.
Let $\{\varphi_t, t \in [0,1]\}$ be a sample - continuous stochastic process. If $\{\varphi_t\}$ is G_t-adapted, we can define the usual Ito integral :

$$\int_0^t \varphi_s \, dW_s = \text{P-lim} \sum_{i=0}^{n-1} \varphi_{t_i^n}(W_{t_{i+1}^n} - W_{t_i^n})$$

where $t_i^n = it/n$.
If $\{\varphi_t\}$is G^t-adapted, using the fact that $\{W_t - W_1, t \in [0,1]\}$ is a "backward G^t-Wiener process", we define the "backward Ito integral":

$$\int_t^1 \varphi_s \oplus dW_s = \text{P-lim} \sum_{i=0}^{n-1} \varphi_{s_{i+1}^n}(W_{s_{i+1}^n} - W_{s_i^n})$$

where $s_i^n = t + i(1-t)/n$. Note that φ is evaluated at the right hand side of the interval (s_i^n, s_{i+1}^n) (we correct here a misprint in [10]), and the resulting integral is a "backward G^t-martingale".

Finally, let us indicate that the convention of summation over repeated indices is used constantly in the sequel.

2. Time reversal of diffusions :

We now state a result, whose proof will appear in [4].
Let $\{X_t, t \in [0,1]\}$ be an $\mathbb{R}^d$-valued process, the solution of the Ito stochastic differential equation :

$$(2.1) \qquad X_t = X_0 + \int_0^t b(s,X_s)ds + \int_0^t \sigma(s,X_s)dW_s$$

where $\{W_t, t \in [0,1]\}$ is an $\mathbb{R}^d$-valued standard Wiener process (standard means that $E(W_t W_t') = t\,I$) defined on a probability space (Ω,F,P), and X_0 is a d-dimensional random vector independent of $\{W_t, t \in [0,1]\}$.

We suppose that the law of X_0 has a bounded density $p(o,x)$, and that $\frac{\partial b_i}{\partial x_j}$, $\frac{\partial \sigma_{ij}}{\partial x_k}$, $\frac{\partial^2 a_{ij}}{\partial x_k \partial x_\ell}$ exist and are bounded functions of (t,x), for $i,j,k,\ell = 1,\ldots,d$; where $a(t,x) = \sigma(t,x)\sigma^*(t,x)$, "$*$" denoting transposed.

Lemma 2.1 : *Under the above hypotheses, $\forall t \in [0,1]$, the law of X_t has a density $p(t,x)$ which satisfies the Fokker-Planck equation :*

$$\text{(2.2)} \qquad \frac{\partial p}{\partial t}(t,x) = \frac{1}{2} \frac{\partial^2}{\partial x_i \partial x_j}(a_{ij}p)(t,x) - \frac{\partial}{\partial x_i}(b_i p)(t,x)$$

and moreover $\exists m \in \mathbb{N}$ s.t.:

$$\int_0^1 \int_{\mathbb{R}^d} \frac{(p^2(t,x) + |(\sigma^* \nabla p)(t,x)|^2)}{1 + |x|^m} \, dx \, dt < \infty$$

□

We want to rewrite $\{X_t, t \in [0,1]\}$ as the solution of a backward stochastic differential equation. For $t \in [0,1]$, define $G^t = \sigma\{W_s - W_t, t \leq s \leq 1\}$ and $H^t = G^t \vee \sigma(X_1)$. We then have also $H^t = G^t \vee \sigma(X_t)$. It is easily seen that $\{W_t - W_1) t \in [0,1]\}$ is a "backward G^t-Wiener process", i.e. $\forall s < t \leq 1$, $W_s - W_t$ is a Gaussian random vector with zero mean and covariance operator $(t-s)I$, which is independent of G^t. The same process is no longer a H^t-Wiener process, but we have :

Theorem 2.2 : *The process $\{\hat{W}_t, t \in [0,1]\}$, whose i-th component is given by :*

$$\hat{W}_t^i = W_t^i - W_1^i - \int_t^1 p(s,X_s)^{-1} \sum_{j=1}^d \frac{\partial}{\partial x_j}(p\sigma_{ji})(s,X_s)ds$$

is a backward H^t-Wiener process (here and in the sequel, any term involving p^{-1} is taken to be zero whenever p is zero).

□

It is now easy to deduce from this result, using the correction between forward and backward Ito's integrals :

Corollary 2.3 : *$\{X_t, t \in [0,1]\}$ solves the following backward Ito stochastic differential equation :*

$$\text{(2.2)} \qquad dX_t = \hat{b}(t,X_t)dt + \sigma(t,X_t) \oplus d\hat{W}_t$$

with final condition the value at $t = 1$ of the solution of (2.1), where

$$\text{(2.3)} \qquad \hat{b}_i(t,x) = b_i(t,x) - (p(t,x))^{-1} \frac{\partial}{\partial x_j}(a_{ij}p)(t,x)$$

3. The forward formulation of the smoothing problem :

We consider now the smoothing problem :

$$(3.1)\qquad \begin{aligned} dX_t &= b(t,X_t)dt + \sigma(t,X_t)dW_t, \ X_0 \text{ given} \\ dY_t &= h(t,X_t)dt + g(t)dW_t + \tilde{g}(t)d\tilde{W}_t, \ Y_0 = 0 \end{aligned}$$

where $\{W_t\}$ and $\{\tilde{W}_t\}$ are independent standard Wiener processes with values in $\mathbb{R}^d$ and $\mathbb{R}^k$ respectively; X_t and Y_t also take values in $\mathbb{R}^d$ and $\mathbb{R}^k$ respectively. Our standing assumptions on b,σ and X_0 are those of section 2. All processes are supposed to be defined on a probability space (Ω,F,P). We now suppose :

(3.2) h is a bounded measurable mapping from $[0,1]\times\mathbb{R}^d$ into $\mathbb{R}^k$.

(3.3) $g(t)g^*(t) + \tilde{g}(t)\tilde{g}^*(t) = I$

The hypothesis that h be bounded is not really necessary for the derivations below, but it seems to be crucial for the proofs of uniqueness to the stochastic partial differential equations we will introduce below. Only in case where $g = 0$ are some uniqueness results for Zakaï equation known for the case of unbounded h (see e.g. PARDOUX [8], OCONE [6]). The hypothesis (3.3) is just a normalization which does simplify the notations below.

We now introduce tmo families of partial differential operators in the variable $x\in\mathbb{R}^d$:

$$L_t = \frac{1}{2}\, a_{ij}(t,x)\frac{\partial^2}{\partial x_i\partial x_j} + b_i(t,x)\frac{\partial}{\partial x_i}$$

$$B_t^\ell = c_{i\ell}(t,x)\frac{\partial}{\partial x_i} + h_\ell(t,x)\,;\ \ell=1,\dots,k$$

where $c(t,x)=\sigma g^*(t,x)$. L_t is the infinitesimal generator of the Markov process $\{X_t\}$; B_t contains the relations between $\{X_t\}$ and $\{Y_t\}$.

Define $F_t^s = \sigma\{Y_u - Y_s\,;\ s\leq u\leq t\}$; and

$Z_t^s = \exp(\int_s^t h_\ell(u,X_u)\, dY_u^\ell - \frac{1}{2}\int_s^t |h(u,X_u)|^2 du)$ where $|.|$ denotes the euclidean norm in $\mathbb{R}^k$. We now introduce a new probability measure $\overset{0}{P}$ on (Ω,F), defined by :

$$\frac{d\overset{0}{P}}{dP} = (\overset{0}{Z_1})^{-1}\ ,\ \text{i.e. } \frac{dP}{d\overset{0}{P}} = \overset{0}{Z_1}$$

It follows from a well-known result on conditional expectation :
$\forall f\in C_b(\mathbb{R}^d)$,

(3.4) $$E[f(X_t)/\overset{0}{F}_1] = \frac{\overset{0}{E}[f(X_t)\overset{0}{Z}_1/\overset{0}{F}_1]}{\overset{0}{E}[\overset{0}{Z}_1 / \overset{0}{F}_1]}$$

where E, $\overset{0}{E}$ denotes expectation with respect to P and $\overset{0}{P}$ respectively. Therefore, if we want to compute the left hand side of (3.4) for all $f \in C_b(\mathbb{R}^d)$, we need only to compute the numerator of the right hand side of (3.4) for all $f \in C_b(\mathbb{R}^d)$. The advantage of doing the computation under $\overset{0}{P}$ is that $\{\overset{0}{F}_t\}$ is the filtration of the $\overset{0}{P}$-Wiener process $\{Y_t, t \in [0,1]\}$ (the latter fact follows from Girsanov's theorem).

We now consider Zakaï's equation :

(3.5) $$d_t q(t,x) = L_t^* q(t,x)dt + B_t^{\ell *} q(t,x)dY_t^\ell ; q_0(x) = p(o,x)$$

By a generalization of our arguments in [7], we obtain :

<u>Proposition 3.1</u> : *Equation (3.4) has a unique solution in the class of $\overset{0}{F}_t$ adapted processes s.t.* $\exists m \in \mathbb{N}$ *with :*

$$E\int_0^1\int_{\mathbb{R}^d} \frac{q^2(t,x)+|(\sigma^*\nabla q)(t,x)|^2}{1+|x|^m}\,dx\,dt < \infty$$

Moreover, $\forall t \in [0,1]$, $\forall f \in C_b(\mathbb{R}^d)$,

$$\int q(t,x)f(x)dx = \overset{0}{E}[f(X_t)\overset{0}{Z}_t / \overset{0}{F}_t]$$

□

We now introduce a backward stochastic partial differential equation :

(3.6) $$d_t v(t,x) + L_t v(t,x)dt + B_t^\ell v(t,x) \oplus dY_t^\ell = 0, v(1,x) = 1.$$

<u>Proposition 3.2</u> : *Equation (3.6) has a unique solution in the class of F_1^t-adapted processes s.t.* $\exists m \in \mathbb{N}$ *with :*

$$E\int_0^1\int_{\mathbb{R}^d} \frac{v^2(t,x)+|(\sigma^*\nabla v)(t,x)|^2}{1+|x|^m}\,dx\,dt < \infty$$

Moreover, $\forall t \in [0,1]$,

$$v(t,x) = \overset{0}{E}_{t,x}[Z_1^t / F_1^t]\ dx\ dP \quad a.e.$$

where $\overset{0}{P}_{t,x}$ is a regular conditional probability distribution of $\{(X_s, Y_s),\ t \leq s \leq 1\}$ under $\overset{0}{P}$, given that $X_t = x$ and $Y_t = o$.

□

It now follows, by the same arguments as those of Theorem 3.7 of [8]:

Theorem 3.3 : *$\forall t \in [0,1]$, $\forall f \in C_b(\mathbb{R}^d)$,*

$$\overset{o}{E}[\,f(X_t)Z_1^0/F_1^0\,] = \int f(x)\; q(t,x)\; v(t,x)dx$$

i.e., in view of formula (3.4),

$q(t,x)v(t,x)[\int_{\mathbb{R}^d} q(t,x)v(t,x)dx]^{-1}$ is the density of the conditional law of X_t, given F_1^0.

□

Remark 3.4 : v(t,x) can be thought of as the Radon-Nikodym derivative of an unnormalized conditional law of X_t given F_1^0 , with respect to an unnormalized conditional law of X_t given F_t^0. We want to point out that, in contradiction with q(t,x),v(t,x) need not be the density of a finite measure. The above solution to the non-linear smoothing problem is very similar to that given by MAYNE [5] and FRASER [2] to the linear smoothing problem, where a "backward filter" is computed with infinite covariance at final time t = 1. However, the difference with the linear theory is that we compute here the conditional density up to a normalization factor. The equation for the normalized filtering density is known. But the Radon-Nikodym derivative of the smoothing law with respect to the filtering law is not F_1^t-adapted, and it is not clear whether one can write an evolution equation for it.

□

4. The backward formulation of the smoothing problem :

In view of Corollary 2.3, we can formulate a backward smoothing problem equivalent to (3.1) (we reverse time in the equation for X_t, replace W_t, in the expression for dY_t, by its expression interms of $\hat{W}_t$, and define $\hat{Y}_t = Y_t - Y_1$):

$$\text{(4.1)}\quad \begin{aligned} dX_t &= \hat{b}(t,X_t)dt + \sigma(t,X_t) \oplus d\hat{W}_t, X_1 \text{ given} \\ d\hat{Y}_t &= \hat{h}(t,X_t)dt + g(t)d\hat{W}_t + \tilde{g}(t)d\tilde{W}_t, \hat{Y}_1 = o \end{aligned}$$

where $\hat{b}$ is given by (2.3), and :

$$\hat{h}_\ell(t,x) = h_\ell(t,x) - (p(t,x))^{-1} \frac{\partial}{\partial x_i}(c_{i\ell}p)(t,x)$$

We now suppose (again this is stronger than what is necessary for what follows):

(4.2) $\hat{h}_\ell$ is a bounded function of t and x; $\ell=1,\ldots,d$.

It follows from HAUSSMANN [3] that a sufficient condition for (4.2) to hold is, in addition to the hypotheses of sections 2 and 3 :

(4.3)(i) $p(o,x) > o$, $\forall x \in \mathbb{R}^d$, $p(o,.)$ is differentiable, and $p(o,x)^{-1}|\nabla p(o,x)|$ is bounded

(4.3)(ii) b^i, as well as all its first and second order partial derivatives in x, is Hölder continuous in (t,x); $\frac{\partial b^i}{\partial x_j \partial x_\ell}(t,x)$ is bounded; $i,j,\ell=1,\ldots,d$.

(4.3)(iii) σ does not depend on x, is Hölder continuous in t, and $\sigma\sigma^*(t) > o$, $\forall t \in [0,1]$.

Note that $\{(\tilde{W}_t, \tilde{W}_t - \tilde{W}_1);\ t \in [0,1]\}$ is a "backward $\overline{H}^t$ Wiener process" and that $(X_t, \hat{Y}_t)$ is adapted to $\overline{H}^t$, where

$$\overline{H}^t = \sigma(W_s - W_1, \tilde{W}_s - W_1;\ t \leqslant s \leqslant 1) \vee \sigma(X_1).$$

We denote again $F_t^s = \sigma(\hat{Y}_u - \hat{Y}_s;\ s \leqslant u \leqslant t) = \sigma(Y_u - Y_s;\ s \leqslant u \leqslant t)$.

Also $dY_t = dY_t$, so that we will drop the "^" over Y_t in the sequel.

We now introduce two families of partial differential operators :

$$\hat{L}_t = \frac{1}{2} a_{ij}(t,x) \frac{\partial^2}{\partial x_i \partial x_j} - \hat{b}_i(t,x) \frac{\partial}{\partial x_i}$$

$$\hat{B}_t^\ell = - c_{i\ell}(t,x) \frac{\partial}{\partial x_i} + \hat{h}_\ell(t,x);\ \ell = 1,\ldots,k$$

Define :

$$\hat{Z}_t^s = \exp\left[\int_s^t \hat{h}_\ell(u,X_u) \oplus dY_u^\ell - \frac{1}{2}\int_s^t |\hat{h}(u,X_u)|^2 du\right];\ o \leqslant s \leqslant t \leqslant 1$$

and a new measure $\overset{1}{P}$ on (Ω,F) given by :

$$\frac{d\overset{1}{P}}{dP} = (\hat{Z}_1^0)^{-1}, \text{ i.e. } \frac{dP}{d\overset{1}{P}} = \hat{Z}_1^0$$

We have again :

$$(4.4)\quad E[f(X_t)/F_1^0] = \frac{\overset{1}{E}[f(X_t)\hat{Z}_1^0/F_1^0]}{\overset{1}{E}[\hat{Z}_1^0/F_1^0]}$$

and $\hat{Y}_t$ is a $\overset{1}{P}$ "backward Wiener process". One cas easily show :

<u>Proposition 4.1</u> : *Let $r(t,x)$ be the unnormalized conditional density of X_t, given F_1^t, i.e. $\forall f \in C_b(\mathbb{R}^d)$,*

$$\overset{1}{E}[f(X_t)\hat{Z}_1^t/F_1^t] = \int_{\mathbb{R}^d} f(x)\, r(t,x)\, dx$$

Then r is a solution (in a weak sense) of the backward Zakaî equation :

$$(4.5)\quad \begin{aligned} d_t r(t,x) + \hat{L}_t r(t,x)dt + \hat{B}_t^{\ell *} r(t,x) \oplus dY_t^\ell = o \\ r(1,x) = p(1,x) \end{aligned}$$

□

Proposition 4.2 : *Let u(t,x) be defined by :*

$$u(t,x) = \overset{1}{E}_{t,x}[\hat{Z}_0^t / F_t^0]$$

Then u is a solution (in a weak sense) of the foward equation :

$$(5.6)\quad \begin{aligned} & d_t u(t,x) = \hat{L}_t u(t,x)dt + \hat{B}_t^\ell u(t,x) dY_t^\ell \\ & u(o,x) = 1 \end{aligned}$$

□

Remark 4.3 : Under the additional assumption (4.3), one can show that (4.5) and (4.6) have unique solutions.

□

We finally have, as in section 3 :

Theorem 4.4 : *$r(t,x)u(t,x)$ is an unnormalized conditional density of X_t given F_1^0, i.e.*

$\forall f \in C_b(\mathbb{R}^d)$,

$$\overset{1}{E}[f(X_t)\hat{Z}_1^0 / F_1^0] = \int f(x) r(t,x) u(t,x) dx$$

□

5. Relations between the forward and backward solutions of the smoothing problem :

Comparing (3.4) and (4.4), we obtain :

$$(5.1)\quad \frac{\overset{0}{E}[f(X_t) Z_1^0 / F_1^0]}{\overset{0}{E}[Z_1^0 / F_1^0]} = \frac{\overset{1}{E}[f(X_t)\hat{Z}_1^0 / F_1^0]}{\overset{1}{E}[\hat{Z}_1^0 / F_1^0]}, \quad \forall f \in C_b(\mathbb{R}^d)$$

We have moreover :

Proposition 5.1 : $\forall t \in [0,1]$,

$q(t,x)\ v(t,x) = r(t,x)\ u(t,x)$ dP dx a.e.

Proof : In view of (5.1), Theorems 3.3 and 4.4, it suffices to show that :

$$\overset{0}{E}[Z_1^0 / F_1^0] = \overset{1}{E}[\hat{Z}_1^0 / F_1^0]$$

We first remark that the restrictions to F_1^0 of $\overset{0}{P}$ and $\overset{1}{P}$ coincide, since both are Wiener measure starting from o at t = o(in the case $g = o$, in fact $\overset{0}{P} = \overset{1}{P}$). Therefore :

$$\overset{0}{E}\left[\frac{d\overset{1}{P}}{d\overset{0}{P}} / F\right] = \frac{d\overset{1}{P}/\overset{0}{F_1}}{d\overset{0}{P}/\overset{0}{F_1}} = 1 \quad \text{a.s.}$$

Using the same formula which led to (3.4) and (4.4), we obtain :

$$\overset{1}{E}[\hat{Z}_1^0 / F_1^0] = \frac{\overset{0}{E}\left[\hat{Z}_1^0 \frac{d\overset{1}{P}}{d\overset{0}{P}} / F_1^0\right]}{\overset{0}{E}\left[\frac{d\overset{1}{P}}{d\overset{0}{P}} / F_1^0\right]}$$

$$= \overset{0}{E}[Z_1^0 / F_1^0] \quad \text{a.s.}$$

since $\dfrac{d\overset{1}{P}}{d\overset{0}{P}} = Z_1^0 \,/\, \hat{Z}_1^0$

□

It now follows :

<u>Theorem 5.2</u> : *$\forall t \in [0,1]$, we have the following identities :*

(i) $q(t,x) = p(t,x)\, u(t,x)$ $\qquad dP\, dx$ a.e.

(ii) $r(t,x) = p(t,x)\, v(t,x)$ $\qquad dP\, dx$ a.e.

<u>Proof</u> : Let us prove (i). We make the particular choice : $h(s,x) = o$, $g(s) = o$ and $\tilde{g}(s) = I$, for $s \in [t,1]$. This choice does not affect the quantities $q(t,x)$ and $u(t,x)$, whereas it does imply that $v(t,x)=1$ and $r(t,x) = p(t,x)$. (i) then follows from Proposition 5.1. (ii) is obtained by a similar argument.

□

<u>Remark 5.3</u> : One can in fact, at least formally, deduce (4.5) and (4.6) from (2.2), (3.5) and (3.6), using the relations (i) and (ii).

□

<u>Corollary 5.4</u> : *The unnormalized conditional density of X_t, given F_1^0 , is given by the four following quantities, which are $dP\, dx$ a.e. equal :*

$$\begin{aligned} q(t,x)\, v(t,x) &= u(t,x)\, r(t,x) \\ &= u(t,x)\, p(t,x)\, v(t,x) \\ &= q(t,x)\, (p(t,x))^{-1} r(t,x) \end{aligned}$$

□

<u>Remark 5.5</u> : The two last expressions obtained in Corollary 5.4 are in a sense more symmetric with respect to time reversal than the two first, except for the factor p(resp. p^{-1}). Indeed, u and v (resp. q and r) are quantities of the same nature. q and r are densities of finite measures, in fact of unnormalized conditional laws (for filtering

problems). u and v are Radon-Nikodym derivatives of the unnormalized smoothing conditional law with respect to unnormalized filtering conditional laws.

The last expression in Corollary 5.4 generalises the result of WALL-WILLSKY-SANDELL [11] in the linear case, where the solution of the linear continuous time smoothing problem is given in terms of two Kalman filters (one forward, one backward) and the a priori law.

□

We have obtained four different expression for the unnormalized conditional density of the fixed-interval non linear smoothing problem. Combining the results in [10] and the formulation of section 2, one can obtain further four other systems of equations for the caracterization of the same conditional law.

REFERENCES

[1] B.D.O. ANDERSON - I.B. RHODES : Smoothing Algorithms for Nonlinear Finite - Dimensional Systems. Stochastics 9, 135-165 (1983)

[2] D.C. FRASER : A New Technique for the Optimal Smoothing of Data, Sc. D. Diss. MIT (1967).

[3] U. HAUSSMANN : On the Drift of a Reversed Diffusion, to appear in Proc. 4th IFIP Conf. on Stoch.Diff.Syst.,M. Métivier - E. Pardoux Eds, Lecture Notes in Control and Information Sciences, Springer-Verlag.

[4] U. HAUSSMANN - E. PARDOUX : Time Reversal of Diffusions. to appear.

[5] D.Q. MAYNE : A Solution of the Smoothing Problem for linear Dynamic Systems, Automatica 4, 73-92 (1966).

[6] D. OCONE : to appear in Proc. 4th IFIP Conf. on Stoch.Diff.Syst., M.Métivier - E.Pardoux Eds, Lecture Notes in Control and Information Sciences, Springer-Verlag.

[7] E. PARDOUX : Stochastic Partial Differential Equations and Filtering of Diffusion Processes. Stochastics 3, 127-167 (1979).

[8] E. PARDOUX : Equations du Filtrage non linéaire, de la Prédiction et du Lissage. Stochastics 6, 193-231 (1982).

[9] E. PARDOUX : Equations of Nonlinear Filtering, and Applications to Stochastic Control with Partial Observation, in Non linear Filtering and Stochastic Control, S. Mitter - A. Moro Eds, Lecture Notes in Mathematics 972, Springer-Verlag (1982).

[10] E. PARDOUX : Equations du Lissage non linéaire. in Filtering and Control of Random Processes, H. Korezlioglu - G. Mazziotto - J. Szpirglas Eds., Lecture Notes in Control and Information Sciences 61, Springer-Verlag (1984).

[11] J. WALL - A. WILLSKY - N. SANDELL : On the Fixed Interval Smoothing Problem. Stochastics 5, 1-41 (1981).

CONSENSUS IN DISTRIBUTED ESTIMATION

P. Varaiya
Department of Electrical Engineering and Computer Sciences
and Electronics Research Laboratory
University of California, Berkeley CA 94720

ABSTRACT

A team must agree on a common decision to minimize the expected cost. Different team members have different observations relating to the 'state of the world', and they may also have different prior beliefs. To reach a consensus they exchange tentative decisions based on their current information. Two questions are discussed: When do the individual estimates converge? If they converge, will a consensus be reached?

1. INTRODUCTION

A team or committee of N people, indexed $i = 1, . . . , N$, must agree on a common decision u to be selected from a pre-specified set U so as to minimize the cost

$$J(\omega, u) \tag{1}$$

where J is a real valued function of the 'state of the world' $\omega \in \Omega$, and the decision u. Initially, different people have different information relating to ω. This is modeled by stipulating that person i observes the value of the random variable $Y_i = Y_i(\omega)$. Everyone knows that i knows Y_i, although $j, j \neq i$, does not know what the value of Y_i actually is. Everyone knows the function J.

Each person has a prior belief concerning ω. We stipulate that i's prior belief is summarized by the probability distribution P^i on (Ω, E) where E is the σ-field of events. If $P^1 = . . = P^N$, we say that the beliefs are *consistent*; otherwise they are *inconsistent*.

Since initially different people have different information, and also because their beliefs may be inconsistent, their estimates of the best decision will also be different. To arrive at a consensus decision it is necessary for them to share information. We suppose that this information is shared by means of the following procedure.

Consider person i. In the first round he makes an estimate $u_i(1)$ which is based on his initial data Y_i, and he communicates this estimate to some or all of the other members. By the time i makes his second estimate, he will have received the estimates of some of the others. More generally, denote by $D_i(t-1)$ the messages received by i from the others before i makes his tth estimate $u_i(t)$. That estimate will be based on Y_i and $D_i(t-1)$. We assume that i communicates all his estimates to a fixed set of the other people, and that there is a message transmission delay of one time unit.

Our aim is to discuss two questions: Will each person's estimate converge as $t \to \infty$? If the individual estimates converge, will they reach a common limit? To formulate these questions mathematically, we need to specify how each person estimates the best decision based on the data available to him. This is done in Section 2. Once this is done, it turns out that the answers depend crucially upon whether the prior beliefs are consistent or inconsitent. The consistent case is considered in Sections 3 and 4, the inconsistent case in Section 5. Section 6 outlines some directions for further research.

2. ESTIMATION SCHEMES

Several different estimation schemes have been considered in the literature.

Borkar and Varaiya [2] consider the situation where the committee wants to estimate a random variable X, and they suppose that the tth estimate made by i, $u_i(t)$, is the conditional mean of X given the available data, i.e.,

$$u_i(t) = E^i \{X \mid Y_i, D_i(t-1)\}. \tag{2}$$

Here E^i denotes expectation with respect to P^i. We will see later that the right hand side of (2) has to be interpreted carefully when the beliefs are inconsistent. For the moment observe that the estimate given by (2) is also the decision that minimizes the (expected value of the) cost function

$$J(\omega, u) := |X(\omega) - u|^2$$

when the information available is $\{Y_i, D_i(t-1)\}$. Aumann [1], and Geanakoplos and Polemarchakis [4] consider the situation in which the group wants to estimate the probability that a particular event $F \in \underline{E}$ has occurred. This is a special case of (2) with $X = 1(F)$.

Tsitsiklis and Athans [7] consider the situation described in the introduction. Sebenius and Geanakoplos [5] discuss the following situation. Suppose there are two people 1, 2, and let $F \in \underline{E}$. Person 1 is allowed to offer a bet, which person 2 may accept or reject. If the bet is accepted and if F is true, i.e., if $\omega \in F$, then 2 must pay 1 a fixed sum of money; whereas if F is false, then 1 must pay 2 the same amount. It follows that 1 will make the bet if and only if the conditional mean of $1(F)$ given the data available to him is ***greater*** than ½, and 2 will accept the bet if and only if the mean conditioned on the data available to 2 is ***less*** than ½. This situation can also be reformulated as a special case of (1) as follows. Take the cost function to be

$$J(\omega, u) := |1(F) - u| \tag{3}$$

and suppose the decision u is restricted to the set $U = \{0, 1\}$. Then person 1 will offer the bet if and only if the decision that minimizes the cost (3) is $u = 1$, and person 2 will accept the bet if and only if the cost minimizing decision is $u = 0$.

Washburn and Teneketzis [8] consider a very general communication framework which includes all the situations mentioned above as special cases. They assume merely that in the tth round, person i selects the decision u according to the rule $u = U_i(Y_i, D_i(t-1))$. The decision rules U_i are known to all.

3. CONSISTENT BELIEFS: I

In this section and the next it is assumed that the beliefs are consistent, i.e.,

$$P^1 = \ldots = P^N := P, \ \textit{say}.$$

Consider first the case studied by Borkar and Varaiya, in which the tth estimate made by i is the mean of X conditioned on $\{Y_i, D_i(t-1)\}$. (It is assumed that $E|X| < \infty$.) Let $\mathcal{Y}_i(t-1)$ denote the σ-field generated by the $\{Y_i, D_i(t-1)\}$. Then

$$u_i(t) = E\,\{X \mid \mathcal{Y}_i(t-1)\}. \tag{4}$$

Let $\mathcal{Y}_i(\infty) := \bigvee_t \mathcal{Y}_i(t)$. Since $\mathcal{Y}_i(t)$ is an increasing sequence, it follows from the martingale convergence thorem that

$$u_i(t) \to u_i(\infty) \ a.s.; \ u_i(\infty) := E\,\{X \mid \mathcal{Y}_i(\infty)\}. \tag{5}$$

Thus the individual estimates do converge.

Next we investigate whether the limiting estimates agree. Suppose i communicates his estimates to j. Then $u_i(t)$ is $\mathcal{Y}_j(t+1)$-measurable. From (5) it follows that $u_i(\infty)$ is $\mathcal{Y}_j(\infty)$-measurable, and so,

$$u_i(\infty) = E\,\{u_j(\infty) \mid \mathcal{Y}_i(\infty) \cap \mathcal{Y}_j(\infty)\}. \tag{6}$$

Suppose there is a *communication ring* $i_1, \ldots, i_n = i_1$. This is a not necessarily distinct sequence of persons such that i_k communicates his estimates to i_{k+1}. Then, according to (6), we must have

$$u_{i_k}(\infty) = E\{u_{i_{k+1}}(\infty) \mid \mathcal{Y}_{i_k}(\infty) \cap \mathcal{Y}_{i_{k+1}}(\infty)\}, \ k = 1, \ldots, n, \tag{7}$$

where $i_{n+1} := i_1$. It is quite easy to show [2, Lemma 2] that (7) implies

$$u_{i_1} = \ldots = u_{i_n},$$

so that the asymptotic estimates of the members of a communication ring agree. This proves the main result of [2]:

THEOREM 1

If the estimates of i are given by (2), then each person's estimate converges. Moreover, if everyone in the team is a member of the same communication ring then the limiting estimates agree.

A careful study of the preceding argument reveals that the estimation scheme (2) enjoys two properties:

(i) The estimate is 'continuous' with respect to a monotonically increasing sequence of data, i.e., (5) holds; and

(ii) The schemes satisfy the 'agreement' property (7).

This observation underlies the work of Washburn and Teneketzis [8] which we discuss next.

4. CONSISTENT BELIEFS: II

The estimate (2) is an instance of a decision rule of the form $u_i(t) = U_i(Y_i, D_i(t-1))$. This suggests the following abstract definitions.

A *decision rule* for i is a function U_i which associates to any σ-field $\mathcal{E}_i \subset \mathcal{E}$ an $\mathcal{E}_i$-measurable random variable $u_i = U_i(\mathcal{E}_i)$ with values in the feasible set U. A decision rule U_i is said to be *continuous* if for every increasing sequence $\mathcal{E}_i(1) \subset \mathcal{E}_i(2) \subset \ldots$, one has $\lim_{k \to \infty} U_i(\mathcal{E}_i(k)) = U_i(\bigvee_k \mathcal{E}_i(k))$ a.s.

Let $U_1, \ldots, U_N$ be a set of decision rules, one for each person. This recursively generates a sequence of estimates by:

$$u_i(t) := U_i(\mathcal{Y}_i(t-1)), \ \mathcal{Y}_i(t-1) := \sigma\{Y_i, D_i(t-1)\}.$$

Let $\mathcal{Y}_i(\infty) := \bigvee_t \mathcal{Y}_i(t)$, and $u_i(\infty) := U_i(\mathcal{Y}_i(\infty))$. The next lemma is immediate from the definition of continuity.

LEMMA 1

If $U_1, \ldots, U_N$ are continuous decision rules, then for every i, $\lim_{t \to \infty} u_i(t) = u_i(\infty)$

a.s.

Suppose $i_1, \ldots, i_n = i_1$ is a communication ring. Then for each k, $u_{i_k}(t)$ is measurable with respect to $\underline{Y}_{i_k}(t) \cap \underline{Y}_{i_{k+1}}(t+1) \subset Y_{i_{k+1}}(\infty)$. If the U_{i_k} are continuous, this yields a chain of conditions:

$$U_{i_k}(\underline{Y}_{i_k}(\infty)) \text{ is } \underline{Y}_{i_{k+1}} - \textit{measurable}, \; k = 1, \ldots, n, \; i_{n+1} = i_1.$$

We say that $U_1, \ldots, U_N$ satisfy the *agreement condition* for *rings* if for every sequence $(i_1, \underline{E}_{i_1}), \ldots, (i_n, \underline{E}_{i_n}) = (i_1, \underline{E}_{i_1})$, the chain of conditions

$$U_{i_k}(\underline{E}_{i_k}) \text{ is } \underline{E}_{i_{k+1}} - \textit{measurable}, \; k = 1, \ldots, n,$$

implies

$$U_{i_1}(\underline{E}_{i_1}) = \ldots = U_{i_n}(\underline{E}_{i_n}).$$

From Lemma 1 and the definition above one can prove the main result of Washburn and Teneketzis.

THEOREM 2

If the decision rules $U_1, \ldots, U_N$ are continuous and satisfy the agreement condition for rings, and if everyone is a member of the same communication ring, then the individual estimates converge to the same limit.

We saw that if the decision rules are simply the conditional means as in (2), then they satisfy the hypothesis of Theorem 2. Washburn and Teneketzis show also that the hypothesis holds when each decision rule is the one that minimizes the expected cost (1), i.e.,

$$U_i(\underline{E}_i) := \textit{arg} \min_{u \in U} E\,\{J(\omega, u) \mid \underline{E}_i\}.$$

Thus Theorem 2 applies to the situation studied by Tsitsitklis and Athans [7].

REMARK

In [2], [3], and [8], the message exchange model is extended to accomodate the situation where (i) i makes a sequence of observations $Y_i(t)$, $t = 1, 2, \ldots$ and not just the initial observation Y_i; and (ii) i communicates his estimate $u_i(t)$ to a randomly selected set of the other people, and message transmission takes a random amount of time. These extensions can easily be incorporated in the analysis and Theorems 1 and 2 continue to hold with some minor modifications.

5. INCONSISTENT BELIEFS

The analysis is quite different when the beliefs are inconsistent. The discussion in this section is based on Teneketzis and Varaiya [6]. To keep the notation simple assume there are only two people, ***Alpha*** and ***Beta***. Initially, ***Alpha*** observes the random variable A and ***Beta*** observes B. Both wish to estimate the random variable X. We also assume that Ω is finite. The prior probabilities of ***Alpha*** and ***Beta*** are denoted P^α, P^β respectively.

For $t = 1, 2, \ldots$ the tth estimate by ***Alpha*** (***Beta***) is denoted α_t (β_t). α_t is the conditional expectation of X given the observations $A, \beta_1, \ldots, \beta_{t-1}$. After α_t has been calculated it is communicated to ***Beta*** whose tth estimate is the conditional expectation of X given $B, \alpha_1, \ldots, \alpha_t$. Once β_t is evaluated it is communicated to ***Alpha*** who incorporates it into the estimate α_{t+1}, and the procedure is repeated.

To complete the specification we assume that the estimation procedures followed by ***Alpha*** and ***Beta*** are consistent with their own prior models. That is, each assumes the other's model to be the same as his own. Consider ***Alpha***. When he receives ***Beta***'s estimate β_{t-1}, ***Alpha*** interprets it as if it were based on P^α rather than on P^β. Thus ***Alpha*** assumes that ***Beta***'s estimate is a realization of the random variable

$$\hat{\beta}_{t-1} := E^\alpha \{X \mid B, \alpha_1, \ldots, \alpha_{t-1}\}.$$

Subsequently, ***Alpha*** calculates α_t,

$$\alpha_t := E^\alpha \{X \mid A, \hat{\beta}_1, \ldots, \hat{\beta}_{t-1}\}.$$

Symmetrically, ***Beta*** interprets α_t as

$$\hat{\alpha}_t := E^\beta \{X \mid A, \beta_1, \ldots, \beta_{t-1}\},$$

and calculates β_t by

$$\beta_t := E^\beta \{X \mid B, \hat{\alpha}_1, \ldots, \hat{\alpha}_t\}.$$

There is a more revealing description of the functional dependence of these estimates. Suppose a particular realization $\bar{\omega} = (\bar{A}, \bar{B})$ has occurred. Since ***Alpha*** observes $\bar{A}$, he concludes that $\bar{\omega} \in \Omega_1^\alpha := \{(A,B) \mid A = \bar{A}\}$ and so his first estimate equals

$$\bar{\alpha}_1 = E^\alpha\{X \mid A = \bar{A}\} = E^\alpha\{X \mid \omega \in \Omega_1^\alpha\}.$$

Alpha transmits the number $\bar{\alpha}_1$ to ***Beta***. ***Beta*** interprets it as a realization of the random variable

$$\hat{\alpha}_1 = E^\beta\{X \mid A\},$$

and so he infers that $\bar{\omega} \in \Omega_1^\beta := \{\omega \mid \hat{\alpha}_1(\omega) = \bar{\alpha}_1, B = \bar{B}\}$, and his first estimate takes the value

$$\overline{\beta}_1 = E^{\beta}\{X \mid \omega \in \Omega_1^{\beta}\}.$$

This value is communicated to *Alpha*.

At the beginning of the tth round, *Alpha* starts with the inference $\overline{\omega} \in \Omega_{t-1}^{\alpha}$ when he receives the estimate $\overline{\beta}_{t-1}$. He interprets it as a realization of the random variable

$$\widehat{\beta}_{t-1} = E^{\alpha}\{X \mid B, \alpha_1, \ldots, \alpha_{t-1}\}$$

and so *Alpha* concludes that $\overline{\omega} \in \Omega_t^{\alpha} := \{\omega \mid \omega \in \Omega_{t-1}^{\alpha}, \widehat{\beta}_{t-1}(\omega) = \overline{\beta}_{t-1}\}$. Hence *Alpha*'s tth estimate takes the value

$$\overline{\alpha}_t = E^{\alpha}\{X \mid \omega \in \Omega_t^{\alpha}\}$$

which is communicated to *Beta*. Whereupon Beta interprets it as a realization of

$$\widehat{\alpha}_t = E^{\beta}\{X \mid A, \beta_1, \ldots, \beta_{t-1}\},$$

concludes that $\overline{\omega} \in \Omega_t^{\beta} := \{\omega \mid \omega \in \Omega_{t-1}^{\beta}, \widehat{\alpha}_t(\omega) = \overline{\alpha}_t\}$, and evaluates his tth estimate as

$$\overline{\beta}_t = E^{\beta}\{X \mid \omega \in \Omega_t^{\beta}\}.$$

Thus, as expected, the uncertainty diminishes with each exchange, $\Omega_{t+1}^{\alpha} \subset \Omega_t^{\alpha}$, $\Omega_{t+1}^{\beta} \subset \Omega_t^{\beta}$. From the description above we also see that if for some k either $\Omega_{k+1}^{\alpha} = \Omega_k^{\alpha}$ or $\Omega_{k+1}^{\beta} = \Omega_k^{\beta}$, then $\Omega_t^{\alpha} = \Omega_{k+1}^{\alpha}$ and $\Omega_t^{\beta} = \Omega_{k+1}^{\beta}$ for $t > k+1$. Hence for $t > T$ (which cannot exceed the number of distinct elements in Ω), Ω_t^{α} and Ω_t^{β} become constant. These limit sets depend upon the realization ω. Call them $\Omega_*^{\alpha}(\omega)$ and $\Omega_*^{\beta}(\omega)$ respectively.

There are two possibilites. The first is that $\Omega_*^{\alpha}(\omega) = \phi$ and $\Omega_*^{\beta}(\omega) = \phi$. This happens because at some stage the message $\overline{\beta}_{t-1}$ received by *Alpha* is "impossible": there is no $\overline{\omega}$ such that $\widehat{\beta}_{t-1}(\overline{\omega}) = \overline{\beta}_{t-1}$; or the message $\overline{\alpha}_t$ received by *Beta* is "impossible": there is no $\overline{\omega}$ such that $\widehat{\alpha}_t(\overline{\omega}) = \overline{\alpha}_t$. *Alpha* and *Beta* must realize that their prior models are inconsistent. Let Ω_I be the set of all realizations that lead to this outcome.

The second possibility is that $\Omega_*^{\alpha}(\omega) \neq \phi$ and $\Omega_*^{\beta}(\omega) \neq \phi$. In this case for $t > T$ the estimates stop changing: $\widehat{\beta}_t(\omega) = \widehat{\beta}_*(\omega)$, $\alpha_t(\omega) = \alpha_*(\omega)$, $\widehat{\alpha}_t(\omega) = \widehat{\alpha}_*(\omega)$, $\beta_t(\omega) = \beta_*(\omega)$. Since for every t, $\widehat{\beta}_t(\omega) = \beta_t(\omega)$ and $\widehat{\alpha}_t(\omega) = \alpha_t(\omega)$, it follows that

$$\widehat{\beta}_*(\omega) = \beta_*(\omega), \quad \widehat{\alpha}_*(\omega) = \alpha_*(\omega).$$

On the other hand, since $\widehat{\beta}_t$ and α_t are based on the same model, namely P^{α}, it follows from Theorem 1 that $\widehat{\beta}_*(\omega) = \alpha_*(\omega)$. For the same reason $\widehat{\alpha}_*(\omega) = \beta_*(\omega)$. Thus if $\omega \in \Omega_{II} := \Omega - \Omega_I$, there is agreement $\alpha_t(\omega) = \beta_t(\omega)$ for $t > T$. It is worth emphasizing that this agreement need not be a reflection of the consistency of the two models P^{α}, P^{β}. Rather agreement occurs because within each person's model there is sufficient "uncertainty" to permit the reconciliation of the other's messages with his own observation. One might say that agreement could result from two wrong arguments. We summarize the preceding analysis as follows.

THEOREM 3

The set of events Ω decomposes into two disjoint subsets Ω_I and Ω_{II}. After T exchanges, if $\omega \in \Omega_I$ both agents realize their models are inconsistent, whereas if $\omega \in \Omega_{II}$ the two estimates coincide.

The result is fragile. In particular, whether a realization ω ends in agreement or in impasse can depend upon the order of communication between *Alpha* and *Beta* as demonstrated in an example in [6].

6. CONCLUDING REMARKS

Recall the discussion in Sections 3, 4. There a consensus is reached via a sequence of exchanges of tentative decisions. The information available to a person increases with each message exchange and the limiting consensus decision is based on the information common to all in the sense that $U_1(\infty) = . . = U_N(\infty)$ is measurable with respect to $\mathcal{Y}_1(\infty) \cap . . \cap \mathcal{Y}_N(\infty)$. A consensus can also be reached if all people share their initial private data $Y_1, . . . , Y_N$. We may call this consensus the ***full information*** decision. It turns out that the consensus reached by exchanging tentative decisions need ***not*** coincide with the full information decision. However, within a rather simple model, Geanakoplos and Polemarchakis [4] have shown that the two decisions are "almost always" the same. It would be worth investigating this in a more general setting.

Secondly, even when the two decisions are the same, it does not follow that all people obtain the full information, i.e., it need not be the case that $\mathcal{Y}_i(\infty) = \sigma\{Y_1, . . . , Y_N\}$. If $\mathcal{Y}_i(\infty)$ is a proper subset of $\sigma\{Y_1, . . . , Y_N\}$, then one could argue that reaching consensus via exchange of tentative decisions requires a transfer of less information than the exchange of all private information. This too is worth further investigation.

Recall now the discussion dealing with the case of inconsistent beliefs. The most interesting finding is that *Alpha* and *Beta* can exchange statements about X and eventually agree even when their views are different. Thus paradoxically, the realization that these views are different is only reached when further communication becomes impossible. This raises several basic and knotty issues that need further investigation.

One can readily imagine situations where the most important thing is to determine whether or not the beliefs are inconsistent. In the communication setup of Section 5 the realization that beliefs are inconsistent is fortuitous -- it happens only if *Alpha* and *Beta* reach an impasse. How should one structure the set of message exchanges so as to expedite the reaching of an impasse?

Suppose now that *Alpha* and *Beta* do reach an impasse ($\omega \in \Omega_I$). Our analysis stops at this point, but there are two directions that can be pursued. First, observe that with the realization that their beliefs are different comes the understanding that they have "misread" each other's messages (i.e. they now know that $\hat{\beta}_n \equiv \beta_n$ and $\hat{\alpha}_n \equiv \alpha_n$), and consequently their estimates have been "biased". To eliminate this bias each needs to learn what the other's view is. A straightforward way of permitting such learning is to suppose that from the beginning *Alpha* admits that *Beta*'s model P^β might be any one of a known set $\underline{P}^\beta$ of models and there is a prior distribution on $\underline{P}^\beta$ reflecting *Alpha*'s initial judgment about *Beta*'s model; a symmetrical structure is formulated for Beta. Within such a framework it seems reasonable to conjecture that each agent will correctly read the other's message and his sequence of estimates will converge. But if their models are different then the limiting estimates may differ, and a consensus will not emerge.

Suppose, however, that *Alpha* and *Beta* want to reach a consensus. To reach a consensus one or both must change their models. One can imagine many different ways in which this can be done. For example, De Groot [3] proposes that each person tells the others what his prior probability is, and he proposes an *ad hoc* behavioral rule whereby each person adjusts his model to a weighted average of the others' models. This is not very satisfactory in situations where communicating one's prior beliefs is not practicable.

ACKNOWLEDGEMENT

This research was supported in part by ONR Contract N00014-80-C-0507 and JSEP Contract F49620-79-C-0178. This paper is based on joint work with Dr. Demos Teneketzis.

REFERENCES

[1] Aumann R J, "Agreeing to disagree," *Annals of Statistics*, vol 4(6), 1976, 1236-1239.

[2] Borkar V and P Varaiya, "Asymptotic agreement in distributed estimation," *IEEE Trans on Automatic Control*, vol AC-27(3), 1982, 650-655.

[3] De Groot M H, "Reaching a consensus," *J American Statistical Association*, Vol 69, No 345, March 1974, 118-121.

[4] Geanakopolos J D and H M Polemarchakis, "We can't disagree forever," *Journal of Economic Theory*, Vol 26, 1982, 192-200.

[5] Sebenius J K and J Geanakoplos, "Don't bet on it: Contingent agreements with asymmetric information," *J American Statistical Association*, Vol 78, No 382, June 1983, 424-426.

[6] Teneketzis D and P Varaiya, "Consensus in distributed estimation with inconsistent beliefs," *Systems and Control Letters*, to appear.

[7] Tsitsiklis J N and M Athans, "Convergence and asymptotic agreement in distributed decision problems," *IEEE Trans on Automatic Control*, vol AC-27(3), 1982, 650-655.

[8] Washburn R B and D Teneketzis, "Asymptotic agreement among communicating decisionmakers," *Stochastics*, to appear. Also, Alphatech Technical Report TR-145, Burlington MA, February 1983.

REMARKS ON LOCAL CONTROLLABILITY OF NONLINEAR CONTROL SYSTEMS†

K. Wagner
Mathematisches Institut
der Universität
Am Hubland
D-8700 Würzburg
GERMANY

1. Introduction

We consider in this paper control systems which are described by ordinary differential equations of the form

$$\dot{x} = f(t,x;u) \ , \ x \in \mathbb{R}^n \ , \ u \in U \subset \mathbb{R}^m \tag{1.1}$$

Here x is the state of the system, and $u = (u^{(1)},\dots,u^{(m)})$ is the control variable. We assume that f is of class C^∞ with respect to (t,x,u). A function $u(t)$ which is piecewise of class C^∞ and assumes its values in U is called an admissible control function. By a solution of (1.1) we mean a pair $u(\cdot)$, $x(\cdot)$ where $u(\cdot)$ is admissible and $x(\cdot)$ is a solution of the differential equation $\dot{x} = f(t,x;u(t))$. $x(\cdot)$ is often called a trajectory of (1.1).

We will assume throughout this paper that there is given a fixed solution $\tilde{u}(\cdot)$, $\tilde{x}(\cdot)$ of (1.1), the "reference solution", defined on some fixed time interval $[t_o,t_e]$. The system (1.1) is called "locally controllable along $\tilde{u}(\cdot)$, $\tilde{x}(\cdot)$ over $[t_o,t_e]$" if, for each $\varepsilon > 0$, a full neighborhood of $\tilde{x}(t_e)$ can be reached, at time t_e, along trajectories $x(\cdot)$ of (1.1) for which $x(t_o) = \tilde{x}(t_o)$ and $\|x(t) - \tilde{x}(t)\| < \varepsilon$ on $[t_o,t_e]$. (1.1) is called "first order controllable along $\tilde{u}(\cdot)$, $\tilde{x}(\cdot)$ over $[t_o,t_e]$" if the variational equation of (1.1) along $\tilde{u}(\cdot)$, $\tilde{x}(\cdot)$ is completely controllable over $[t_o,t_e]$ (cf. /5/, Sec. 3). By variational equation we mean as usual the linear system

$$\begin{aligned} \dot{y} &= A(t)\cdot y + B(t)\cdot u \\ A(t) &:= f_x(t,\tilde{x}(t);\tilde{u}(t)) \ , \ B(t) := f_u(t,\tilde{x}(t);\tilde{u}(t)) \end{aligned} \tag{1.2}$$

It is clear that "first order controllability" implies "local controllability". Note also that both properties refer to "large" time intervals - in contrast to the notion of (small time) local controllabi-

†This research constitutes a part of the author's Ph.D. dissertation. It was supported by Deutsche Forschungsgemeinschaft Kn 164/2-1.

lity along a reference trajectory which is commonly used in the literature (see e.g. /4/).

We continue in this paper the discussions started in /5/ which concern the relation between local controllability for a given system (1.1) and the corresponding property for some associated system. If the underlying system (1.1) is of the affine type, i.e. if it can be written as

$$\dot{x} = f(t,x;u) = f_o(t,x) + \sum_{\nu=1}^{m} g_\nu(t,x)u^{(\nu)} \qquad (1.3)$$

then the associated system is a formal extension using elements from the Lie algebra generated by $g_1,\ldots,g_m$ (with the usual Lie bracket product $[h,k] = k_x h - h_x k$). That is, one selects suitable elements g_ν, $\nu > m$, from this Lie algebra and adds them to the right hand side of (1.3) after scalar multiplication with new control variables $u^{(\nu)}$. This kind of comparison technique started with the classical work of Chow /1/ and was carried further in /2/, /3/, /5/. As a typical result we repeat here the following theorem (Thm. 6.1 in /5/):

Besides (1.3), consider the associated system

$$\dot{x} = f_o(t,x) + \sum_{\nu=1}^{m} g_\nu(t,x)u^{(\nu)} + \sum_{1\leq\mu<\nu\leq m} [g_\nu,g_\mu](t,x)u^{(\nu,\mu)} \qquad (1.4)$$

Let there be given a C^∞ solution $\tilde{u}(\cdot)$, $\tilde{x}(\cdot)$ of (1.3), defined on $[t_o,t_e]$, and consider it as a solution of (1.4) with $\tilde{u}^{(\nu,\mu)} = 0$. Suppose that (1.4) is first order controllable over $[t_o,t_e]$ along this solution. Then, for each $\varepsilon > 0$, one can find a solution of (1.3) along which (1.3) is first order controllable over $[t_o,t_e]$ and whose trajectory $\tilde{x}_r(\cdot)$ satisfies $\tilde{x}_r(t) = \tilde{x}(t)$ for $t = t_o,t_e$ and $\|\tilde{x}_r(t) - \tilde{x}(t)\| \leq \varepsilon$ on $[t_o,t_e]$. (However, in general, $\tilde{x}_r(\cdot)$ differs from $\tilde{x}(\cdot)$).

In this paper we undertake the attempt to generalize this result to a nonaffine system of the form (1.1). A major difference to the affine case concerns the notion of an associated system. This system is linear with respect to the control and therefore no formal extension of (1.1). In explicit terms, it is given as

$$\begin{aligned}\dot{x} &= f(t,x;\tilde{u}(t)) + \sum_{\nu=1}^{m} b^\nu(t,x)u^{(\nu)} + \sum_{1\leq\mu<\nu\leq m} [b^\nu,b^\mu](t,x)u^{(\nu,\mu)} \\ b^\nu(t,x) &:= (\partial f/\partial u^{(\nu)})(t,x;\tilde{u}(t))\end{aligned} \qquad (1.5)$$

Note that the reference trajectory $\tilde{x}(\cdot)$ of (1.1) is also a trajectory of (1.5), however, in general, trajectories of (1.1) will not be trajectories of (1.5) and vice versa - an essential difference to the affine case. Our main result then runs as follows: If (1.5) is first order controllable along $u \equiv 0$, $x = \tilde{x}(\cdot)$ over $[t_o,t_e]$ (and if an extra condition holds, see below) then (1.1) is first order controllable along appropriate neighboring solutions (as described in the theorem quoted above).

It should be noted that one cannot deduce this result from /5/, Thm. 6.1 just by linearizing with respect to the controls. The reason is that, from this theorem, one well obtains first order controllability of the system

$$\dot{x} = f(t,x;\tilde{u}(t)) + \sum_{\nu=1}^{m} b^{\nu}(t,x)u^{(\nu)} \tag{1.6}$$

but along a solution which is, in general, different from $u \equiv 0$, $x = \tilde{x}(\cdot)$ and which therefore does not arise from a solution of (1.1) via linearization.

A further difference between the situation considered in this paper and the previous one is the appearance of an extra condition. It concerns the vectors

$$(\partial^2 f/\partial u^{(\nu)}\partial u^{(\mu)})(t,x;\tilde{u}(t)) \; , \; 1 \leqq \mu,\nu \leqq m \tag{1.7}$$

which do not vanish identically unless the system is affine. In order to prove our main theorem we require that all elements (1.7) are contained (in a sense which will be made precise, see (2.10)) in a linear space which we denote by $\mathcal{L}(t,x)$. In the affine case this space is, roughly speaking, the subspace generated by those Lie-elements which arise from the g_ν by repeated application of the ad operator. The precise definition in the general case is given in Sec. 2.

Finally, a word on the structure of the paper: In the sections 2 and 3 we adopt the techniques developed in /5/, Sec. 2 and 4, for the nonaffine case. The results of Sec. 2 which are relevant for the following parts are summed up in Lemma 2.3. Section 4 contains the statement and the proof of our generalization of /5/, Thm. 6.1. In the last section we indicate without proof further controllability criteria which are related to equality type necessary conditions for optimal solutions (/6/, Sec. 21). The link between these results and the ones presented here is the similarity in the techniques employed here and in /6/.

2. Auxiliary results

The subject of the following considerations is a control system of the form (1.1), together with some fixed reference solution $\tilde{u}(\cdot)$, $\tilde{x}(\cdot)$, defined on $[t_o,t_e]$, $t_o < t_e$. f is supposed to be a C^∞ function of (t,x,u), defined on a set $W \times U$ where W is a neighborhood of $\{(t,x) : t_o \leqq t \leqq t_e, \|x - \tilde{x}(t)\| \leqq r\}$, $r > 0$. We assume that $\tilde{u}(\cdot)$ is a

C^∞ function of t and that $\tilde{u}(t)$ belongs to the interior of U for each $t \in [t_o, t_e]$. By admissible control functions we mean functions of t (eventually also of further parameters) which take their values in U and which are of class C^∞ with respect to all their arguments, except for jumps at finitely many hyperplanes $t = t_i$.

Throughout this section, we make the following additional assumption which is of a technical nature and will be dropped later:

$$\begin{aligned} &f(t,x;0) = 0 \quad \text{identically in } (t,x) \in W \\ &\tilde{x}(\cdot) \equiv 0 \ , \ \tilde{u}(\cdot) \equiv 0 \end{aligned} \tag{2.1}$$

The main technical tool we shall use in this section is the quantity $C(t,x,z_1,\ldots,z_N;\mathbf{o},\mathbf{u}_1,\ldots,\mathbf{u}_N)$ associated with (1.1) which has been introduced in /6/, Sec. 8. We shortly recall some properties of $C(\cdot)$: $C(\cdot)$ is a n-valued formal power series in the N scalar variables $z_1, \ldots, z_N$, each of whose coefficients is a n-valued C^∞ function depending on t, x and finitely many out of a collection of m-dimensional vectors $u_{i,j}$, $1 \leq i \leq N$, $j = 0,1,2,\ldots$ $\mathbf{u}_i$ is an abbreviation for the sequence $(u_{i,o}, u_{i,1}, u_{i,2}, \ldots)$. Roughly spoken, the series $C(\cdot)$ describes the effect of some special kind of control variation applied near the point (t,x) and defined in terms of the z_i, $\mathbf{u}_i$, $i = 1,\ldots,N$ (cf. Lemma 8.1, /6/, and Lemma 2.4, /5/).

We wish to study in this section in some detail the lower order terms of $C(\cdot)$. To this purpose, we recall some definitions of further quantities introduced in /6/. Let $\mathbf{u} = (u_o, u_1, \ldots)$ be a sequence of m-valued independent variables and $g(t,x,\mathbf{u})$ be a C^∞ function depending on t, x and finitely many elements u_i. Then the operator Γ is defined as

$$\Gamma g := \partial g/\partial t + \sum_{j=o}^{\infty} (\partial g/\partial u_j)\cdot u_{j+1} + [f,g] \tag{2.2}$$

(cf. /6/, p.72, eq.(13.1); here $f(t,x,\mathbf{u}) := f(t,x;u_o)$). Furthermore, we introduce the following matrix valued functions:

$$\begin{aligned} &B_o(t,x,\mathbf{u}) := (\partial f/\partial u)(t,x;u_o) \\ &B_{\nu+1}(t,x,\mathbf{u}) := (\Gamma B_\nu)(t,x,\mathbf{u}) \ , \ \nu = 0,1,2,\ldots \end{aligned} \tag{2.3}$$

(cf. /6/, p.75, eq.(13.7)). Here the application of Γ to the $m \times n$ matrix B_ν has to be understood as its application to each of its columns. The ρ-th column of B_ν will be denoted by B_ν^ρ. The space $\mathcal{L}(t,x)$ mentioned in the introduction can now be described exactly: it is the linear span of $\{B_\nu^\rho(t,x,\tilde{\mathbf{u}}(t)) : \rho = 1,\ldots,m, \ \nu = 0,1,2,\ldots\}$, where $\tilde{\mathbf{u}}(t) = (\tilde{u}(t), \dot{\tilde{u}}(t), \ddot{\tilde{u}}(t), \ldots)$. For later purposes we note the following rule: if $\alpha = \alpha(t,x)$ is a m-valued C^∞ function on W then

$$(\Gamma(B_\nu\cdot\alpha))(t,x,\mathbf{o}) = (B_{\nu+1}\cdot\alpha)(t,x,\mathbf{o}) + (B_\nu\cdot\beta)(t,x,\mathbf{o}) \tag{2.4}$$

where $\mathbf{o} = (0,0,0,\dots)$ and $\beta = \beta(t,x)$ is an appropriate m-valued C^∞ function depending on α (cf. /6/, Lemma 13.1).

Lemma 2.1. Let d be a fixed positive integer. Let

$$\mathbf{u}_o = \mathbf{o}\ ,\ \tilde{\mathbf{u}}_i = (\tilde{u}_{i,o}, \tilde{u}_{i,1}, \dots) := \mathbf{u}_{i-1} - \mathbf{u}_i\ ,\ i = 1,\dots,N \tag{2.5}$$

Then, for the system (1.1), the following asymptotic formula holds (the arguments $(t,x,\mathbf{o})$ of the B_ν and the derivatives of f are omitted):

$$\begin{aligned} &C(t,x,z_1,\dots,z_N;\mathbf{o},\mathbf{u}_1,\dots,\mathbf{u}_N) = \\ &= x + \frac{1}{d!}\sum_{j=1}^{d-1} B_{d-1-j}\cdot\binom{d-1}{j}\sum_{i=1}^{N}\tilde{u}_{i,j}z_i^d + \\ &+ \frac{1}{(d+1)!}\sum_{j=1}^{d} B_{d-j}\cdot\binom{d}{j}\sum_{i=1}^{N}\tilde{u}_{i,j}z_i^{d+1} + \\ &+ \sum_{1\leq\mu\leq\nu\leq m}(\partial^2 f/\partial u^{(\nu)}\partial u^{(\mu)})\cdot(p_{2,1}^{(\nu,\mu)}(u_{i,o};z_i) + p_{1,1,2}^{(\nu,\mu)}(u_{i,o};u_{i,1};z_i)) + \\ &+ \sum_{1\leq\mu\leq\nu\leq m}\Gamma(\partial^2 f/\partial u^{(\nu)}\partial u^{(\mu)})\cdot p_{2,2}^{(\nu,\mu)}(u_{i,o};z_i) \\ &+ \frac{1}{2}\sum_{1\leq\mu<\nu\leq m}[B_o^\nu,B_o^\mu]\cdot\sum_{1\leq i<j\leq N} z_j(z_i - z_j)(\tilde{u}_{i,o}^{(\nu)}\tilde{u}_{j,o}^{(\mu)} - \tilde{u}_{i,o}^{(\mu)}\tilde{u}_{j,o}^{(\nu)}) + \\ &+ \mathcal{O}(\|z\|\sum_{i=1}^{N}\|\mathbf{u}_i\|^3 + \|z\|^3\sum_{i=1}^{N}\|\mathbf{u}_i\|^2 + \|z\|^{d+2}\sum_{i=1}^{N}\|\mathbf{u}_i\| + \\ &\qquad + \sum_{1\leq j\leq\nu<d}\|\sum_{i=1}^{N}\tilde{u}_{i,j}z_i^\nu\| + \sum_{\nu=1}^{d+1}\|\sum_{i=1}^{N}\tilde{u}_{i,o}z_i^\nu\|) \end{aligned} \tag{2.6}$$

Here the $p_{2,1}^{(\nu,\mu)}$, $p_{1,1,2}^{(\nu,\mu)}$, $p_{2,2}^{(\nu,\mu)}$ are polynomials with constant coefficients in the $u_{i,o}$, $u_{i,1}$, z_i which are homogeneous with respect to the N-tuples $(u_{1,o},\dots,u_{N,o})$, $(u_{1,1},\dots,u_{N,1})$, $(z_1,\dots,z_N)$. The corresponding degrees are indicated in the subscripts (e.g. each $p_{2,1}^{(\nu,\mu)}(u_{i,o};z_i)$ is homogeneous of degree 2 with respect to $(u_{1,o},\dots,u_{N,o})$ and of degree 1 w.r.t. $(z_1,\dots,z_N)$).

Remark. If f is linear with respect to u and if $d = 1$ then the statement of Lemma 2.1 can be found in /5/, Lemma 2.1. Note that the scalar factors appearing before the $[B_o^\nu,B_o^\mu]$ are exactly the same for the linear and the nonlinear case. The only difference is that, in the nonlinear case, extra terms appear which involve second order derivatives of f with respect to u.

Proof. Since terms which are of order ≥ 3 with respect to $\mathbf{u}$ are of no interest, one can, for the purpose of the proof, replace C by the second order approximation $C^{(2)}$. $C^{(2)}$ is defined and described explicitely in /6/, eqs. (14.3), (14.6) which run as follows (note that $\mathbf{u}_o = \mathbf{o}$ in our case):

$$\begin{aligned} C^{(2)} &= x + \sum_{\nu=1}^{d+1} \frac{1}{\nu!} \cdot \sum_{i=1}^{N} l^{(\nu)}(\mathbf{o},\tilde{\mathbf{u}}_i) z_i^{\nu} + \\ &+ \frac{1}{2} \sum_{\nu=1}^{2} \frac{1}{\nu!} \cdot \sum_{i=1}^{N} L^{(\nu)}(\mathbf{o},\tilde{\mathbf{u}}_i) \cdot (\mathbf{u}_{i-1} + \mathbf{u}_i) z_i^{\nu} - \\ &- \frac{1}{2} \sum_{1 \leq i < j \leq N} [l^{(1)}(\mathbf{o},\tilde{\mathbf{u}}_j), l^{(1)}(\mathbf{o},\tilde{\mathbf{u}}_i)] z_j z_i + \\ &+ O(\|z\|^{d+2} \sum_{i=1}^{N} \|\mathbf{u}_i\| + \|z\|^3 \sum_{i=1}^{N} \|\mathbf{u}_i\|^2 + \|\sum_{i=1}^{N} \tilde{u}_{i,o} z_i\|) \end{aligned} \tag{2.7}$$

To explain the general meaning of $l^{(\nu)}$, $L^{(\nu)}$ we introduce sequences $\mathbf{V} = (v_o, v_1, \ldots)$, $\mathbf{W} = (w_o, w_1, \ldots)$ of m-valued independent variables, different from the $\mathbf{u}_i$. According to /6/, Theorem 13.1, we have

$$l^{(\nu)}(t,x;\mathbf{u},\mathbf{V}) = \sum_{j=o}^{\nu-1} \binom{\nu-1}{j} B_{\nu-1-j}(t,x,\mathbf{u}) \cdot v_j \;, \quad \nu = 1,2,3,\ldots \tag{2.8}$$

Using the following identities in $(t,x,\mathbf{u})$

$$\frac{\partial B_1^{\mu}}{\partial u_o^{(\nu)}} = \Gamma\left(\frac{\partial B_o^{\mu}}{\partial u_o^{(\nu)}}\right) + [B_o^{\nu}, B_o^{\mu}], \quad \frac{\partial B_1^{\mu}}{\partial u_1^{(\nu)}} = \frac{\partial B_o^{\mu}}{\partial u_o^{(\nu)}} = \frac{\partial^2 f}{\partial u^{(\nu)} \partial u^{(\mu)}}$$

(see/6/, Lemmas 13.1 and 13.2) we get from (2.8) and /6/, Def. 12.1:

$$\begin{aligned} L^{(1)}(t,x;\mathbf{u},\mathbf{V}) \cdot \mathbf{W} &= \sum_{\nu,\mu=1}^{m} \frac{\partial^2 f(t,x;u_o)}{\partial u^{(\nu)} \partial u^{(\mu)}} \cdot v_o^{(\mu)} w_o^{(\nu)} \\ L^{(2)}(t,x;\mathbf{u},\mathbf{V}) \cdot \mathbf{W} &= \sum_{\nu,\mu=1}^{m} \frac{\partial^2 f(t,x;u_o)}{\partial u^{(\nu)} \partial u^{(\mu)}} \cdot (v_1^{(\mu)} w_o^{(\nu)} + v_o^{(\mu)} w_1^{(\nu)}) + \\ &+ \sum_{\nu,\mu=1}^{m} \left(\Gamma\left(\frac{\partial B_o^{\mu}}{\partial u_o^{(\nu)}}\right) + [B_o^{\nu}, B_o^{\mu}]\right)(t,x,\mathbf{u}) \cdot v_o^{(\mu)} w_o^{(\nu)} \end{aligned} \tag{2.9}$$

Substituting (2.9) and (2.8) into (2.7) yields the desired formula for $C^{(2)}$ except for the $[B_o^{\nu}, B_o^{\mu}]$ part which runs as follows:

$$\begin{aligned} &\frac{1}{2} \cdot \frac{1}{2!} \cdot \sum_{\nu,\mu=1}^{m} [B_o^{\nu}, B_o^{\mu}](t,x,\mathbf{o}) \cdot \sum_{j=1}^{N} \tilde{u}_{j,o}^{(\mu)} (u_{j-1,o}^{(\nu)} + u_{j,o}^{(\nu)}) z_j^2 - \\ &- \frac{1}{2} \cdot \sum_{\nu,\mu=1}^{m} [B_o^{\nu}, B_o^{\mu}](t,x,\mathbf{o}) \cdot \sum_{1 \leq i < j \leq N} \tilde{u}_{j,o}^{(\nu)} \tilde{u}_{i,o}^{(\mu)} z_j z_i \end{aligned}$$

That this can be rewritten as

$$\frac{1}{2} \cdot \sum_{1 \leq \mu < \nu \leq m} [B_o^{\nu}, B_o^{\mu}](t,x,\mathbf{o}) \cdot \sum_{1 \leq i < j \leq N} z_j (z_i - z_j) (\tilde{u}_{j,o}^{(\mu)} \tilde{u}_{i,o}^{(\nu)} - \tilde{u}_{i,o}^{(\mu)} \tilde{u}_{j,o}^{(\nu)})$$

has been demonstrated in /5/ (cf. eqs. (2.5) and (2.6)). Therefore the proof is completed. □

We next wish to generalize the Lemma 2.2, /5/, to general nonaffine control systems. There we will need for the first time the crucial condition concerning the vectors (1.7) which roughly says that the derivatives $\partial^2 f/\partial u^{(\nu)}\partial u^{(\mu)}$ are contained in the linear span of the B_σ, not only along the reference solution but in a whole neighborhood of its trajectory in the (t,x)-space. As in /5/, we denote by $v^{(\nu)}$, $1 \leq \nu \leq m$, $v^{(\nu,\mu)}$, $1 \leq \mu < \nu \leq m$, a family of independent variables which, as a whole, will be abbreviated by $\underline{v}$.

Lemma 2.2. Assume that there exist $d \in \mathbb{N}$ and m-valued C^∞ functions $\alpha^{(j)}_{\nu,\mu}(t,x)$ such that for all $(t,x) \in W$

$$\frac{\partial^2 f}{\partial u^{(\nu)}\partial u^{(\mu)}}(t,x;0) = \sum_{j=0}^{d-3} B_j(t,x,\mathbf{o})\cdot\alpha^{(j)}_{\nu,\mu}(t,x)\ , \quad 1\leq\mu\leq\nu\leq m \tag{2.10}$$

Claim: There exist a positive integer N, real numbers $-1 < z_1 < z_2 < \ldots < z_N < 0$ and sequences $\mathbf{u}_i(t,x,\underline{v}) = (u_{i,0}(\cdot), u_{i,1}(\cdot),\ldots)$ such that the following two statements hold true for all $(t,x) \in W$ and all $\underline{v}$:

1. $u_{i,j}(\cdot) \equiv 0$ for $j > d$. For $j \leq d$, the components of $u_{i,j}(\cdot)$ are polynomials in $\underline{v}$ whose coefficients are C^∞ functions of (t.x).
2. Let λ be a real parameter. Then

$$C(t,x,\lambda z;\mathbf{o},\lambda^{d-1}\mathbf{u}_1(t,x,\underline{v}),\ldots,\lambda^{d-1}\mathbf{u}_N(t,x,\underline{v})) =$$

$$= x + \lambda^{2d}\Big(\sum_{\nu=1}^{m} B_0^\nu(t,x,\mathbf{o})v^{(\nu)} + \sum_{1\leq\mu<\nu\leq m}[B_0^\nu,B_0^\mu](t,x,\mathbf{o})v^{(\nu,\mu)}\Big) + \tag{2.11}$$
$$+\ \mathcal{O}(\lambda^{2d+1}).$$

Proof. According to the rule (2.4) it follows from (2.10) that there also exist m-valued C^∞ functions $\beta^{(j)}_{\nu,\mu}(t,x)$ such that for $(t,x) \in W$

$$\Gamma\Big(\frac{\partial^2 f}{\partial u^{(\nu)}\partial u^{(\mu)}}\Big)(t,x,\mathbf{o}) = \sum_{j=0}^{d-2} B_j(t,x,\mathbf{o})\cdot\beta^{(j)}_{\nu,\mu}(t,x) \tag{2.12}$$

Now insert (2.10) and (2.12) into the expansion (2.6), substitute $z \to \lambda z$, $\mathbf{u}_i \to \lambda^{d-1}\mathbf{u}_i$ and compare coefficients in (2.6) and (2.11). Right from the beginning we put $u_{i,j} = 0$ for $j > d$ and also $u_{i,1} = 0$ (therefore all $p^{(\nu,\mu)}_{1,1,2}(u_{i,0};u_{i,1};z_i)$ in (2.6) vanish). We first wish to determine $\tilde{u}_{i,0}$ such that

$$\frac{1}{2}\sum_{1\leq i<j\leq N} z_j(z_i - z_j)(\tilde{u}^{(\nu)}_{i,0}\tilde{u}^{(\mu)}_{j,0} - \tilde{u}^{(\mu)}_{i,0}\tilde{u}^{(\nu)}_{j,0}) = v^{(\nu,\mu)},\quad 1\leq\mu<\nu\leq m \tag{2.13}$$

$$\sum_{i=1}^{N}\tilde{u}_{i,0}z_i^\sigma = 0 \quad \text{for } \sigma = 1,\ldots,d+1 \tag{2.14}$$

Except for the fact that (2.14) has to hold not only for $\sigma = 1,2$ the conditions (2.13), (2.14) coincide with the system of equations (2.8), (2.12) in /5/. It has been shown in /5/, Lemma 2.3, that this system can be solved, for each sufficiently large N and each fixed choice of numbers $-1 < z_1 < z_2 < \dots < z_N < 0$, by polynomials $\tilde{u}_{i,o}(\underline{v})$ with constant coefficients. The same argument will work here - just by increasing (if necessary) the number N if the additional linear conditions $\sigma = 3,\dots,d+1$ have to be taken into account.

We now can insert the constructed $\tilde{u}_{i,o}(\underline{v})$ into the right hand side of (2.6). By comparing coefficients one sees easily that, in order to complete the proof, one only has to solve, separately for each $j = 2,\dots,d$, a system of linear equations for the $\tilde{u}_{i,j}$, $i = 1,\dots,N$. The right hand sides of these systems are polynomials in $\underline{v}$ whose coefficients are C^∞ functions of (t,x). The coefficient matrix of each system is constant and coincides, in essential, with a collection of $d+1$ rows of the Vandermonde matrix emerging from $z_1,\dots,z_N$, provided we have chosen $N > d+1$. Therefore it is clear that, for these systems, there exist solutions of the form as stated in the lemma. □

We now are in the position to state the analogue of Lemma 2.4 in /5/. The proof is almost a repetition of the one given in /5/ and is therefore omitted (one only has to use our Lemma 2.2 instead of Lemma 2.2 in /5/).

Lemma 2.3. Consider the control system (1.1), fulfilling the extra condition (2.1). Assume in addition that the hypothesis (2.10) of Lemma 2.2 is satisfied. Then there exists a n-valued function $\tilde{c}(t,x;\underline{v},\lambda)$ (where λ denotes a real parameter) which has the following properties:

1. For each choice of a bounded set $\underline{V}$ in the $\underline{v}$-space, $\tilde{c}$ is of class C^∞ on a neighborhood of a set

   $\{t,x,\underline{v},\lambda : t \in [t_o,t_e],\ \|x\| \leq r,\ \underline{v} \in \underline{V},\ |\lambda| \leq \lambda_o\}$

   where $\lambda_o > 0$ depends on the choice of $\underline{V}$.
2. $\tilde{c}$ admits the following asymptotic expansion:

$$\tilde{c}(t,x;\underline{v},\lambda) = x + \lambda^{2d}\Big(\sum_{\nu=1}^{m} B_o^\nu(t,x,\circ)v^{(\nu)} + \sum_{1\leq\mu<\nu\leq m} [B_o^\nu, B_o^\mu](t,x,\circ)v^{(\nu,\mu)}\Big) + \mathcal{O}(\lambda^{2d+1}) \tag{2.15}$$

   which holds uniformly with respect to t, x, $\underline{v}$ restricted on compact sets.
3. To each fixed values $\tilde{t} \in [t_o,t_e]$, $\lambda \in [0,\lambda_o]$ there exists an admissible control function $u(t;\tilde{x},\underline{v})$, $t \in [t_o,t_e]$, $\|\tilde{x}\| \leq r$,

$\underline{v} \in \underline{V}$, such that

a) $\|u(t;\tilde{x},\underline{v})\| = \mathcal{O}(\lambda^{d-1})$ (2.16)

$u(t;\tilde{x},\underline{v}) = 0$ if $t \notin [\tilde{t} + \lambda z_1, \tilde{t}]$, where $z_1 \in (-1,0)$ is a constant independent of $\tilde{t}$, $\tilde{x}$, $\underline{v}$, λ.

b) The solution $x(t;\tilde{x})$ of the initial value problem

$\dot{x} = f(t,x;u(t;\tilde{x},\underline{v}))$, $x(t_o') = \tilde{x}$

(here $t_o' \in [t_o, \tilde{t} + \lambda z_1]$ arbitrary) satisfies

$$x(\tilde{t};\tilde{x}) = \tilde{c}(\tilde{t},\tilde{x};\underline{v},\lambda) \tag{2.17}$$

$$\|x(t;\tilde{x})\| = \mathcal{O}(\|\tilde{x}\| + \lambda^d) \tag{2.18}$$

(2.16), (2.18) hold uniformly in $t \in [t_o', \tilde{t}]$, $t_o', \tilde{t} \in [t_o, t_e]$, $\underline{v} \in \underline{V}$, $\|\tilde{x}\| \leq r$.

3. Transformation of the State

Consider a general system of the form (1.1). From now on we do no longer assume that (2.1) holds. Given a fixed reference solution $\tilde{u}(\cdot)$, $\tilde{x}(\cdot)$ which is of class C^∞, let $x(t,\bar{x})$ be the general solution of

$\dot{x} = f(t,x;\tilde{u}(t))$, $x(t_e) = \tilde{x}(t_e) + \bar{x}$

$x(t,\bar{x})$ is defined and of class C^∞ on some neighborhood $\bar{W}$ of $\{(t,\bar{x}): t_o \leq t \leq t_e, \ \bar{x} = 0\}$. As is well known, for $(t,\bar{x}) \in \bar{W}$, the Jacobian $X(t,\bar{x}) := (\partial x/\partial\bar{x})(t,\bar{x})$ exists and is nonsingular. In analogy to /5/, Sec. 4, we now wish to use $x(t,\bar{x})$ in order to define a transformation of the system (1.1) which is t-dependent and in general nonlinear. We write the transformed system as

$$\dot{\bar{x}} = \bar{f}(t,\bar{x};\bar{u}) \tag{3.1}$$

where

$$\bar{f}(t,\bar{x};\bar{u}) := X(t,\bar{x})^{-1}\{f(t,x(t,\bar{x});\bar{u} + \tilde{u}(t)) - f(t,x(t,\bar{x});\tilde{u}(t))\} \tag{3.2}$$

Note that if $\bar{u}(\cdot)$, $\bar{x}(\cdot)$ is a solution of (3.1) then $u(t) := \bar{u}(t) + \tilde{u}(t)$, $x(t) := x(t,\bar{x}(t))$ is a solution of (1.1) and vice versa. The given reference solution $\tilde{u}(\cdot)$, $\tilde{x}(\cdot)$ corresponds to the stationary solution $\tilde{\bar{u}} \equiv 0$, $\tilde{\bar{x}} \equiv 0$ of (3.1), and the assumption (2.1) holds true for the transformed system.

We need in the sequel the following formulas which relate the significant quantities of the original and the transformed systems (we again use a bar to denote the latter ones):

$$\bar{B}_\sigma(t,\bar{x},\bar{\mathfrak{v}}) = X(t,\bar{x})^{-1} \cdot B_\sigma(t,x(t,\bar{x}),\bar{\mathfrak{v}} + \tilde{\mathfrak{v}}(t)) \ , \ \sigma = 0,1,2,\ldots \tag{3.3}$$

where $\tilde{\mathfrak{v}}(t) := (\tilde{u}(t), \dot{\tilde{u}}(t), \ddot{\tilde{u}}(t),\ldots)$

$$\frac{\partial^2 \bar{f}}{\partial \bar{u}^{(\nu)} \partial \bar{u}^{(\mu)}}(t,\bar{x};\bar{u}) = X(t,\bar{x})^{-1} \frac{\partial^2 f}{\partial u^{(\nu)} \partial u^{(\mu)}}(t,x(t,\bar{x});\bar{u}+\tilde{u}(t)) \qquad (3.4)$$

The proof of these formulas consists in a straightforward computation using the definitions (2.2), (2.3), (3.2); therefore we shall omit it here.

4. The Main Result

We now come to the formulation of our main result: the generalization of Theorem 6.1, /5/, to control systems of the form (1.1) in which the control u enters nonlinearly.

We consider the same situation as described at the beginning of Section 2. In particular, we assume that there is given a fixed reference solution $\tilde{u}(\cdot)$, $\tilde{x}(\cdot)$ of (1.1), defined on $[t_o, t_e]$. The appropriate associated system we have to look at besides (1.1) is then given by the equation

$$\dot{x} = f(t,x;\tilde{u}(t)) + \sum_{\nu=1}^{m} B_o^{\nu}(t,x,\tilde{\mathbf{u}}(t))u^{(\nu)} + \\ + \sum_{1 \leq \mu < \nu \leq m} [B_o^{\nu}, B_o^{\mu}](t,x,\tilde{\mathbf{u}}(t))u^{(\nu,\mu)} \qquad (4.1)$$

Here $\tilde{\mathbf{u}}(t) = (\tilde{u}(t), \dot{\tilde{u}}(t), \ddot{\tilde{u}}(t), \ldots)$, and $u^{(\nu)}$, $u^{(\nu,\mu)}$ are the new control variables which are allowed to assume arbitrary real values. Obviously, the given reference trajectory $\tilde{x}(t)$ together with the zero control constitutes a solution of (4.1) on $[t_o, t_e]$.

Theorem 4.1. Assume that the system (4.1) is first order controllable along the solution $x \equiv \tilde{x}(t)$, $u \equiv 0$ over $[t_o, t_e]$. Assume further that there exists a positive integer d and m-valued C^{∞} functions $\alpha_{\nu,\mu}^{(j)}(t,x)$ such that, identically in $(t,x) \in W$,

$$\frac{\partial^2 f}{\partial u^{(\nu)} \partial u^{(\mu)}}(t,x;\tilde{u}(t)) = \sum_{j=o}^{d-3} B_j(t,x,\tilde{\mathbf{u}}(t)) \cdot \alpha_{\nu,\mu}^{(j)}(t,x) \qquad (4.2)$$

Claim: Given any $\varepsilon > 0$, there exists a solution $\tilde{u}_r(t)$, $\tilde{x}_r(t)$, $t_o \leq t \leq t_e$, of the system (1.1) such that

1. $\|\tilde{x}_r(t) - \tilde{x}(t)\| \leq \varepsilon$ for $t_o \leq t \leq t_e$. $\tilde{x}_r(t) = \tilde{x}(t)$ for $t = t_o, t_e$.
   $\|\tilde{u}_r(t) - \tilde{u}(t)\| \leq \varepsilon$ for $t_o \leq t \leq t_e$.
2. The system (1.1) is first order controllable along $\tilde{x}_r(\cdot)$, $\tilde{u}_r(\cdot)$ over the interval $[t_o, t_e]$.

Proof. In essential, the proof follows the same pattern as the one for Theorem 6.1 in /5/. Therefore we can be rather brief here and restrict ourselves to carry out those parts of the proof in detail in which digressions from /5/ actually occur.

The hypothesis on the system (4.1) is equivalent to the following statement (cf. /5/, Sec. 3): There exists an admissible control function $\underline{u}(t,p)$ with components $u^{(\nu)}(t,p)$, $u^{(\nu,\mu)}(t,p)$ which are linear and homogeneous with respect to the n-dimensional parameter p, such that the solution $\underline{x}(t,p)$ of the initial value problem

$$\dot{x} = f(t,x;\tilde{u}(t)) + \sum_{\nu=1}^{m} B_o^\nu(t,x,\tilde{u}(t))u^{(\nu)}(t,p) + \\ + \sum_{1\leq\mu<\nu\leq m} [B_o^\nu, B_o^\mu](t,x,\tilde{u}(t))u^{(\nu,\mu)}(t,p) \ , \quad x(t_o) = \tilde{x}(t_o)$$

satisfies $\underline{x}(t_e,p) = \tilde{x}(t_e) + p + \mathcal{O}(\|p\|^2)$.

Now we apply the state transformation described in Sec. 3 to both systems (1.1) and (4.1). It is easy to see that the transformed systems are related to each other in the same way as the original systems (1.1) and (4.1) are, and that (4.1) after the transformation assumes the form

$$\dot{x} = \sum_{\nu=1}^{m} B_o^\nu(t,x,0)u^{(\nu)} + \sum_{1\leq\mu<\nu\leq m} [B_o^\nu, B_o^\mu](t,x,0)u^{(\nu,\mu)} \tag{4.3}$$

So without loss of generality we can work in the sequel with these additional hypotheses:

(i) $f(t,x;0) = 0$ for $(t,x) \in W$, $\tilde{u}(\cdot) \equiv 0$, $\tilde{x}(\cdot) \equiv 0$ (i.e. the condition (2.1) holds true),

(ii) there exists an admissible control $\underline{u}(t,p)$ with components $u^{(\nu)}(\cdot)$, $u^{(\nu,\mu)}(\cdot)$ such that the solution of

$$\dot{x} = \sum_{\nu=1}^{m} B_o^\nu(t,x,0)u^{(\nu)}(t,p) + \sum_{1\leq\mu<\nu\leq m} [B_o^\nu, B_o^\mu](t,x,0)u^{(\nu,\mu)}(t,p)$$

$$x(t_o) = 0$$

satisfies the terminal condition $x(t_e) = p + \mathcal{O}(\|p\|^2)$.

As a consequence of these conditions we now can apply the results of Sec. 2. We use the function $\tilde{c}$ constructed in Lemma 2.3 to introduce the new function

$$c(t,x;p,\lambda) := \frac{1}{\lambda}(\tilde{c}(t,x;\underline{u}(t,p),\lambda) - x) \ , \quad \lambda > 0 \tag{4.4}$$

(i.e. we made the substitution $\underline{v} \to \underline{u}(t,p)$). Using statement 2 of Lemma 2.3 and the homogeneity of $\underline{u}(t,p)$ with respect to p we get

$$c(t,x;p,\lambda) = \sum_{\nu=1}^{m} B_o^\nu(t,x,0)u^{(\nu)}(t,\lambda^{2d-1}p) + \\ + \sum_{1\leq\mu<\nu\leq m} [B_o^\nu, B_o^\mu](t,x,0)u^{(\nu,\mu)}(t,\lambda^{2d-1}p) + \mathcal{O}(\lambda^{2d}) \\ = \mathcal{O}(\lambda^{2d-1}). \tag{4.5}$$

By the same type of argument as used in /5/ one concludes from (4.5) that the solution $x(t,p)$ of the initial value problem

$$\dot{x} = c(t,x;p,\lambda) \ , \ x(t_o) = 0 \tag{4.6}$$

satisfies

$$\|x(t,p)\| = \mathcal{O}(\lambda^{2d-1}) \ , \ x(t_e,p) = \lambda^{2d-1}p + \mathcal{O}(\lambda^{2d}) \tag{4.7}$$

provided p is bounded and λ is sufficiently small.

The next step is the approximation of the trajectories of (4.6) on $[t_o,t_e]$ by the Euler-Cauchy type polygons as described in /5/, Sec. 5. So we put $\lambda = (t_e - t_o)/J$, J some sufficiently large integer,

$$\begin{aligned} &\xi_o(p) := 0 \ , \ \text{and for } j = 1,\dots,J: \\ &t_j := t_{j-1} + \lambda \\ &\xi_j(p) := \tilde{c}(t_j,\xi_{j-1}(p);\underline{u}(t_j,p),\lambda) = c(t_j,\xi_{j-1}(p);p,\lambda) \end{aligned} \tag{4.8}$$

At this point we remark that the function $c(t,x;p,\lambda)$ is, in general, only piecewise of class C^∞ with respect to t. The discontinuities arise at those t where $\underline{u}(t,p)$ becomes discontinuous. However, since these t do not depend on x, p, λ, it can be easily shown that the error estimates for the procedure (4.8) given in /5/, Lemma 5.1, remain valid. Therefore one can conclude from (4.5), (4.7) that, in analogy to /5/, eq.(6.13), the following statement holds: if λ is chosen sufficiently small, say $\lambda \leqq \lambda_o$, and if $J = (t_e - t_o)/\lambda$ is an integer then

$$\xi_j(p) = \mathcal{O}(\lambda^{2d-1}) \ , \ j = 1,\dots,J \tag{4.9}$$

$$t_J = t_e \ , \ \xi_J(p(q)) = \lambda^{2d-1}q \tag{4.10}$$

Here $p(q)$ is an appropriate C^∞ function of q depending on the choice of λ and defined on some ball $\|q\| \leqq \pi$. $\pi > 0$ here depends only on λ_o, not on λ. Furthermore, for fixed λ, each $\xi_j(p)$, $j = 0,\dots,J$, is a C^∞ function of p. This follows from (4.8) by an induction argument.

The final step now consists in connecting the discrete points t_j, $\xi_j(p)$, $j = 0,\dots,J$, by means of trajectories of the system $\dot{x} = f(t,x;u)$. Applying statement 3 of Lemma 2.3 one infers from (4.8) that, for fixed λ, there exists for each $j = 1,\dots,J$ an admissible control $u_j(t;x,p)$ such that $\xi_j(p)$ is the terminal value at $t = t_j$ of the solution of the initial value problem

$$\dot{x} = f(t,x;u_j(t;\xi_{j-1}(p),p)) \ , \ x(t_j - \lambda) = \xi_{j-1}(p)$$

Furthermore, we deduce from (4.9) and (2.18) that this solution is of order $\mathcal{O}(\lambda^d)$, uniformly with respect to $t \in [t_o,t_e]$ and $p \in \{p(q): \|q\| \leqq \pi\}$. We therefore can confine the corresponding trajectory to a prescribed ε-neighborhood of our reference trajectory $\tilde{x}(\cdot) \equiv 0$ by choosing λ sufficiently small. In order to get $\|u_j(\cdot)\| < \varepsilon$ we can proceed in the same way because of (2.16) (of course we may assume that $d \geqq 2$). From now on, λ is regarded as fixed.

Since each $\xi_j(p)$ is a C^∞ function of p the definition $u(t,p) := u_j(t;\xi_{j-1}(p),p)$, respectively, for $t \in (t_o + (j-1)\lambda, t_o + j\lambda]$, respectively, yields an admissible control function which steers the system $\dot{x} = f(t,x;u)$ from t_o, 0 to t_e, $\xi_J(p)$. Therefore, using as control for the original system the function $u(t,p(q))$ we arrive at the terminal point $\lambda^{2d-1}q$. q now can be regarded as a n-dimensional parameter varying in a neighborhood of $q = 0$, and $u(t,p(q))$, regarded as a function of t and q, is admissible. The last statement therefore implies first order controllability of the original system (1.1) over $[t_o,t_e]$ along the trajectory through t_o, $\tilde{x}(t_o)$ generated by $u(t,p(0))$. □

5. Further Extensions

In this section we state - without going into any details of the proofs - further criteria for local controllability of the same type as Theorem 4.1. The common background of these results are the various types of higher order necessary conditions for optimal solutions which have been derived in /6/. In fact, the subsequent propositions have the same structure as the basic three first and second order equality type conditions, i.e. those conditions which assume the form of an orthogonality relation $y^T q = 0$, y the adjoint vector. Again, as in Theorem 4.1, we have to strengthen the hypothesis used in /6/: certain relations have to hold not only along the reference solution but in a full neighborhood of its trajectory. The technique is a modification of the one used in /6/. What one actually has to do is to create a situation analogous to that underlying Lemma 2.2, and then proceed in precisely the same way as before. The formal background again is taken from /6/, however some extra work is necessary to adopt this technique for our purposes.

The first proposition is a consequence of /6/, Lemma 14.2.

Proposition 5.1. Suppose that the hypotheses as specified at the beginning of Section 2 are satisfied. Then the assertion of Theorem 4.1 holds if, for some $\rho \geqq 0$, the system

$$\dot{x} = f(t,x;\tilde{u}(t)) + \sum_{\sigma=o}^{\rho} \sum_{\nu=1}^{m} B_\sigma^\nu(t,x,\tilde{u}(t))u^{(\sigma,\nu)} \tag{5.1}$$

(with control variables $u^{(\sigma,\nu)}$) is first order controllable along $u \equiv 0$, $x = \tilde{x}(\cdot)$ over the interval $[t_o,t_e]$.

The following result can be obtained by using the special formal series C which appeared in /6/ in the proofs of the theorems 21.1 and 21.2. We again consider the system (1.1) but we assume now in particular that the control dimension m is equal to 2.

Proposition 5.2. Suppose that the hypotheses as specified at the beginning of Section 2 are satisfied. The assertion of Theorem 4.1 holds if one of the following conditions I or II is fulfilled:

I. (i) For some positive integer d, the vectors

$$\begin{aligned} &(\partial B_o^1/\partial u_o^{(1)})(t,x,\tilde{u}(t)) \\ &[B_{\sigma-1}^1, B_\sigma^1](t,x,\tilde{u}(t)), \quad \sigma = 1,\dots,\rho, \ \rho \geqq 0 \end{aligned} \tag{5.2}$$

are, identically in $(t,x) \in W$, linear combinations of the columns of $B_j(t,x,\tilde{u}(t))$, $j = 0,\dots,d-1$, with C^∞ coefficients.

(ii) The system

$$\dot{x} = f(t,x;\tilde{u}(t)) + \sum_{\nu=1}^{2} B_o^\nu(t,x,\tilde{u}(t))u^{(\nu)} + \frac{\partial B_\rho^2}{\partial u_o^{(1)}}(t,x,\tilde{u}(t))u^{(3)} \tag{5.3}$$

(which has the control variables $u^{(1)}$, $u^{(2)}$, $u^{(3)}$) is first order controllable along $u \equiv 0$, $x = \tilde{x}(t)$ over $[t_o, t_e]$.

II. (i) For some positive integer d, the vectors

$$\begin{aligned} &(\partial B_o^i/\partial u_o^{(i)})(t,x,\tilde{u}(t)) \\ &[B_{\sigma-1}^i, B_\sigma^i](t,x,\tilde{u}(t)), \quad \sigma = 1,\dots,\rho_i, \ \rho_1 \geqq \rho_2 \geqq 0 \end{aligned} \quad (i = 1,2) \tag{5.4}$$

are, identically in $(t,x) \in W$, C^∞ linear combinations of the columns of $B_j(t,x,\tilde{u}(t))$, $j = 0,\dots,d-1$.

(ii) The system

$$\dot{x} = f(t,x;\tilde{u}(t)) + \sum_{\nu=1}^{2} B_o^\nu(t,x,\tilde{u}(t))u^{(\nu)} + \frac{\partial B_{\rho_1+\rho_2+1}^2}{\partial u_o^{(1)}}(t,x,\tilde{u}(t))u^{(3)} \tag{5.5}$$

is first order controllable along $u \equiv 0$, $x = \tilde{x}(t)$ over $[t_o, t_e]$.

References

/1/ Chow, W.L., Über Systeme von linearen partiellen Differentialgleichungen 1. Ordnung, Math. Annalen 117 (1940/41), 98-105

/2/ Hermann, R., On the Accessibility Problem in Control Theory, in: International Symposium on Nonlinear Differential Equations, LaSalle, Lefschetz eds., Academic Press, London 1963

/3/ Haynes, G.W., Hermes, H., Nonlinear Controllability via Lie Theory, SIAM J. Control 8 (1970), 450-460

/4/ Sussmann, H.J., Lie Brackets and Local Controllability: A Sufficient Condition for Scalar-Input Systems, SIAM J. Control 21 (1983), 686-713

/5/ Knobloch, H.W., Wagner, K., On Local Controllability of Nonlinear Systems, in: Dynamical Systems and Microphysics, G. Leitmann ed., Academic Press, to appear

/6/ Knobloch, H.W., Higher Order Necessary Conditions in Optimal Control Theory, Lecture Notes in Control and Information Sciences Vol. 34, Springer, Berlin 1981

Lecture Notes in Control and Information Sciences

Edited by A. V. Balakrishnan and M. Thoma

Vol. 43: Stochastic Differential Systems
Proceedings of the 2nd Bad Honnef Conference of the SFB 72 of the DFG at the University of Bonn
June 28 – July 2, 1982
Edited by M. Kohlmann and N. Christopeit
XII, 377 pages. 1982.

Vol. 44: Analysis and Optimization of Systems
Proceedings of the Fifth International Conference on Analysis and Optimization of Systems
Versailles. December 14–17, 1982
Edited by A. Bensoussan and J. L. Lions
XV, 987 pages, 1982

Vol. 45: M. Arató
Linear Stochastic Systems with Constant Coefficients
A Statistical Approach
IX, 309 pages. 1982

Vol. 46: Time-Scale Modeling of Dynamic Networks with Applications to Power Systems
Edited by J. H. Chow
X, 218 pages. 1982

Vol. 47: P. A. Ioannou, P. V. Kokotovic
Adaptive Systems with Reduced Models
V, 162 pages. 1983

Vol. 48: Yaakov Yavin
Feedback Strategies for Partially Observable Stochastic Systems
VI, 233 pages, 1983

Vol. 49: Theory and Application of Random Fields
Proceedings of the IFIP-WG 7/1 Working Conference held under the joint auspices of the Indian Statistical Institute
Bangalore, India, January 1982
Edited by G. Kallianpur
VI. 290 pages. 1983

Vol. 50: M. Papageorgiou
Applications of Automatic Control Concepts to Traffic Flow Modeling and Control
IX, 186 pages. 1983

Vol. 51: Z. Nahorski, H. F. Ravn, R. V. V. Vidal
Optimization of Discrete Time Systems
The Upper Boundary Approach
V, 137 pages 1983

Vol. 52: A. L. Dontchev
Perturbations, Approximations and Sensitivity Analysis of Optimal Control Systems
IV, 158 pages. 1983

Vol. 53: Liu Chen Hui
General Decoupling Theory of Multivariable Process Control Systems
XI, 474 pages. 1983

Vol. 54: Control Theory for Distributed Parameter Systems and Applications
Edited by F. Kappel, K. Kunisch, W. Schappacher
VII, 245 pages. 1983.

Vol. 55: Ganti Prasada Rao
Piecewise Constant Orthogonal Functions and Their Application to Systems and Control
VII, 254 pages. 1983.

Vol. 56: Dines Chandra Saha, Ganti Prasada Rao
Identification of Continuous Dynamical Systems
The Poisson Moment Functional (PMF) Approach
IX, 158 pages. 1983.

Vol. 57: T. Söderström, P. G. Stoica
Instrumental Variable Methods for System Identification
VII, 243 pages. 1983.

Vol. 58: Mathematical Theory of Networks and Systems
Proceedings of the MTNS-83 International Symposium
Beer Sheva, Israel, June 20–24, 1983
Edited by P. A. Fuhrmann
X, 906 pages. 1984

Vol. 59: System Modelling and Optimization
Proceedings of the 11th IFIP Conference
Copenhagen, Denmark, July 25-29, 1983
Edited by P. Thoft-Christensen
IX, 892 pages. 1984

Vol. 60: Modelling and Performance Evaluation Methodology
Proceedings of the International Seminar
Paris, France, January 24–26, 1983
Edited by F. Bacelli and G. Fayolle
VII, 655 pages. 1984

Vol. 61: Filtering and Control of Random Processes
Proceedings of the E.N.S.T.-C.N.E.T. Colloquium
Paris, France, February 23–24, 1983
Edited by H. Korezlioglu, G. Mazziotto, and J. Szpirglas
V, 325 pages. 1984